Geometry

Instruction Manual

By Steven P. Demme

1-888-854-MATH (6284)

www.MathUSee.com

Math·U·See

1-888-854-MATH (6284)
www.MathUSee.com

Graphic Design by Christine Minnich
Illustrations by Gregory Snader

Printed in the United States of America

Geometry

SCOPE AND SEQUENCE
HONORS TOPICS

HOW TO USE
SUPPORT AND RESOURCES

LESSON   1   Points, Lines, Rays, and Line Segments
LESSON   2   Planes and Sets
LESSON   3   Angles
LESSON   4   Types of Angles
LESSON   5   Parallel and Perpendicular Lines
LESSON   6   Supplementary and Complementary Angles
LESSON   7   Transversals
LESSON   8   Perimeter; Interior Angles
LESSON   9   Area
LESSON 10   Constructing and Identifying Triangles

LESSON 11   Regular Polygons
LESSON 12   Geometry of a Circle, Sphere, and Ellipse
LESSON 13   Area of a Circle and an Ellipse
LESSON 14   Volume of Rectangular Solid and Cylinder
LESSON 15   Volume: Pyramid, Cone, Prism, and Sphere
LESSON 16   Surface Area of Solids
LESSON 17   Radicals
LESSON 18   Pythagorean Theorem
LESSON 19   More on Radicals
LESSON 20   Special Triangles: 45°- 45°- 90°

LESSON 21   Special Triangles: 30°- 60°- 90°
LESSON 22   Axioms, Postulates, and Theorems
LESSON 23   Corresponding Parts of Triangles
LESSON 24   Proving Triangles Congruent: SSS and SAS
LESSON 25   Proving Triangles Congruent: ASA and AAS
LESSON 26   Proving Right Triangles Congruent
LESSON 27   Proving Triangles Similar with AA
LESSON 28   Transformational Geometry
LESSON 29   Trigonometric Functions
LESSON 30   Inverse Trigonometric Functions

STUDENT SOLUTIONS
HONORS SOLUTIONS
TEST SOLUTIONS

SYMBOLS & TABLES
GEOMETRY GLOSSARY
SECONDARY LEVELS MASTER INDEX
ALGEBRA REVIEW TOPICS
GEOMETRY INDEX

Math·U·See

SCOPE & SEQUENCE

Math-U-See is a complete and comprehensive K-12 math curriculum. While each book focuses on a specific theme, Math-U-See continuously reviews and integrates topics and concepts presented in previous levels.

Primer

α Alpha | Focus: Single-Digit Addition and Subtraction

β Beta | Focus: Multiple-Digit Addition and Subtraction

γ Gamma | Focus: Multiplication

δ Delta | Focus: Division

ε Epsilon | Focus: Fractions

ζ Zeta | Focus: Decimals and Percents

Pre-Algebra

Algebra 1

Stewardship*

Geometry

Algebra 2

Pre Calculus With Trigonometry

Calculus

*Stewardship is a biblical approach to personal finance. The requisite knowledge for this curriculum is a mastery of the four basic operations, as well as fractions, decimals, and percents. In the Math-U-See sequence these topics are thoroughly covered in Alpha through Zeta. We also recommend Pre-Algebra and Algebra 1 since over half of the lessons require some knowledge of algebra. Stewardship may be studied as a one-year math course or in conjunction with any of the secondary math levels.

HONORS TOPICS

Here are the topics for the special challenge lessons included in the student text. You will find one honors page after the last systematic review page for each regular lesson. Instructions for the honors pages are included in the student text.

LESSON	TOPIC
01	Solving logic problems using charts
02	Union and intersection with Venn diagrams; Mobius strip
03	Using three-part Venn diagrams to solve word problems
04	Using a compass rose; word problems requiring equations
05	Forming patterns by bisecting lines and angles
06	More patterns formed by bisection; word problems
07	Using known information to find out about other lines and angles
08	Using known information to find out sizes of angles in a geometric drawing
09	Challenge area problems; doubling and squaring dimensions
10	Golden rectangle
11	More logic problems
12	Advanced construction—equilateral triangle, regular octagon, regular hexagon in a circle
13	Finding the area of irregular polygons on a grid of dots
14	Finding the area of any triangle when three sides are known; Relationships between dimensions and volume
15	Word problems involving volume
16	Visualization of the formula for the area of a circle; Finding different volumes with the same surface area
17	Archimedes and the relationship between surface area and volume of spheres and cylinders
18	Using Pythagorean theorem to find sight distance from a height
19	Volume and surface area of oblique prism; word problems
20	Simple vectors using a protractor to measure angles
21	Finding the area of a ring by measuring the chord of the circle
22	Introduction to the language of formal logic
23	The converse in formal logic
24	Using remote interior angles and other information to find angle measures
25	Geometric proofs
26	More geometric proofs
27	Proofs involving circles, tangents, and chords
28	Proofs with circles and inscribed angles
29	Applications using tangent function
30	Applications using sine and cosine functions

HOW TO USE

Five Minutes for Success

Welcome to *Geometry*. I believe you will have a positive experience with the unique Math-U-See approach to teaching math. These first few pages explain the essence of this methodology, which has worked for thousands of students and teachers. I hope you will take five minutes and read through these steps carefully.

I am assuming that your student has a thorough grasp of the four basic operations (addition, subtraction, multiplication, and division), along with a mastery of fractions, decimals, percents, pre-algebra, and algebra 1.

If you are using the program properly and still need additional help, you may contact your authorized representative or visit Math·U·See online at http://www.mathusee.com/support.html. —**S. Demme**

The Goal of Math-U-See

The underlying assumption or premise of Math-U-See is that the reason we study math is to apply math in everyday situations. Our goal is to help produce confident problem solvers who enjoy the study of math. These are students who learn their math facts, rules, and formulas *and* are able to use this knowledge in solving word problems and real-life applications. Therefore, the study of math is much more than simply committing to memory a list of facts. It includes memorization, but it also encompasses learning underlying concepts that are critical to problem solving.

More than Memorization

Many people confuse memorization with understanding. Once while I was teaching seven junior high students, I asked how many pieces they would each receive if there were fourteen pieces. The students' response was, "What do we do: add, subtract, multiply, or divide?" Knowing *how* to divide is important; understanding *when* to divide is equally important.

THE SUGGESTED 4-STEP MATH-U-SEE APPROACH

In order to train students to be confident problem solvers, here are the four steps that I suggest you use to get the most from the Math-U-See curriculum:

Step 1. Preparation for the lesson.
Step 2. Presentation of the new topic.
Step 3. Practice for mastery.
Step 4. Progression after mastery.

Step 1. Preparation for the lesson.

Watch the DVD to learn the concept. Study the written explanations and examples in the instruction manual. Many students watch the DVD along with their instructor. Older students in the secondary level who have taken responsibility to study math themselves will do well to watch the DVD and read through the instruction manual.

For this book, you will need a straightedge or ruler, a protractor, and a compass for measuring angles. I do not encourage calculators for most work at this level. However, they may be useful for a few concepts. These are noted in the lessons.

Step 2. Presentation of the new topic.

Now that you have studied the new topic, choose problems from the first lesson practice page to present the new concept to your students.

   a. **Build:** The geometry course does not ask you to build with the manipulatives. However, the student will benefit from drawing or sketching many of the problems.

   b. **Write:** Record the step-by-step solutions on paper as you work them.

   c. **Say:** Explain the "why" and "what" of math as you build and write.

Do as many problems as you feel are necessary until the student is comfortable with the new material. One of the joys of teaching is hearing a student say, *"Now I get it!"* or *"Now I see it!"*

Step 3. Practice for mastery.

Using the examples and the lesson practice problems from the student text, have the students practice the new concept until they understand it. It is one thing for students to watch someone else do a problem, it is quite another to do the same problem themselves. Do enough examples together so that they can do them without assistance.

Do as many of the lesson practice pages as necessary (not all pages may be needed) until the students remember the new material and gain understanding. Give special attention to the word problems, which are designed to apply the concept being taught in the lesson.

Step 4. Progression after mastery.

Once mastery of the new concept is demonstrated, proceed into the systematic review pages for that lesson. Mastery can be demonstrated by having each student teach the new material back to you. The goal is not to fill in worksheets, but to be able to teach back what has been learned.

The systematic review worksheets review the new material as well as provide practice of the math concepts previously studied. Remediate missed problems as they arise to ensure continued mastery.

After the last systematic review page in each lesson, you will find an "honors" lesson. These are optional, but highly recommended for students who will be taking advanced math or science courses. These challenging problems are a good way for all students to hone their problem-solving skills.

Proceed to the lesson tests. These were designed to be an assessment tool to help determine mastery, but they may also be used as extra worksheets. Your students will be ready for the next lesson only after demonstrating mastery of the new concept and continued mastery of concepts found in the systematic review worksheets.

Confucius is reputed to have said, "Tell me, I forget; show me, I understand; let me do it, I will remember." To which we add, **"Let me teach it and I will have achieved mastery!"**

Length of a Lesson

So how long should a lesson take? This will vary from student to student and from topic to topic. You may spend a day on a new topic, or you may spend several days. There are so many factors that influence this process that it is impossible to predict the length of time from one lesson to another. I have spent three days on a lesson, and I have also invested three weeks in a lesson. This occurred in the same book with the same student. If you move from lesson to lesson too quickly without the student demonstrating mastery, he will become overwhelmed and discouraged as he is exposed to more new material without having learned the previous topics. But if you move too slowly, your student may become bored and lose interest in math. I believe that as you regularly spend time working along with your student, you will sense when is the right time to take the lesson test and progress through the book.

By following the four steps outlined above, you will have a much greater opportunity to succeed. Math must be taught sequentially, as it builds line upon line and precept upon precept on previously learned material. I hope you will try this methodology and move at your student's pace. As you do, I think you will be helping to create a confident problem solver who enjoys the study of math.

ONGOING SUPPORT AND ADDITIONAL RESOURCES

Welcome to the Math-U-See Family!

Now that you have invested in your children's education, I would like to tell you about the resources that are available to you. Allow me to introduce you to your regional representative, our ever improving website, the Math-U-See blog, our new free e-mail newsletter, the online Forum, and the Users Group.

Most of our regional **Representatives** have been with us for over 10 years. What makes them unique is their desire to serve and their expertise. They have all used Math-U-See and are able to answer most of your questions, place your student(s) in the appropriate level, and provide knowledgeable support throughout the school year. They are wonderful!

Come to your local curriculum fair where you can meet your rep face-to-face, see the latest products, attend a workshop, meet other MUS users at the booth, and be refreshed. We are at most curriculum fairs and events. To find the fair nearest you, click on "Events Calendar" under "News."

The **Website**, at www.mathusee.com, is continually being updated and improved. It has many excellent tools to enhance your teaching and provide more practice for your student(s).

Math-U-See Blog

Interesting insights and up-to-date information appear regularly on the Math-U-See Blog. The blog features updates, rep highlights, fun pictures, and stories from other users. Visit us and get the latest scoop on what is happening.

Email Newsletter

For the latest news and practical teaching tips, sign up online for the free Math-U-See e-mail newsletter. Each month you will receive an e-mail with a teaching tip from Steve as well as the latest news from the website. It's short, beneficial, and fun. Sign up today!

The Math-U-See Forum and the Users Group put the combined wisdom of several thousand of your peers with years of teaching experience at your disposal.

Online Forum

Have a question, a great idea, or just want to chitchat with other Math-U-See users? Go to the online forum. You can also use the forum to post a specific math question if you are having difficulty in a certain lesson. Head on over to the forum and join in the discussion.

Yahoo Users Group

The MUS-users group was started in 1998 for lovers and users of the Math-U-See program. It was founded by two home-educating mothers and users of Math-U-See. The backbone of information and support is provided by several thousand fellow MUS users.

Online Solutions

If you detect a questionable answer, go to http://www.mathusee.com/solutions.php for the most up-to-date solutions. If the answer is not accurate, please send the correction to corrections@mathusee.com.

For Specific Math Help

When you have watched the DVD instruction and read the instruction manual and still have a question, we are here to help. Call your local rep, click the support link and e-mail us here at the home office, or post your question on the forum. Our trained staff have used Math-U-See themselves and are available to answer a question or walk you through a specific lesson.

Feedback

Send us an e-mail by clicking the feedback link. We are here to serve you and help you teach math. Ask a question, leave a comment, or tell us how you and your student are doing with Math-U-See.

Our hope and prayer is that you and your students will be equipped to have a successful experience with math!

Blessings,

Steve

Steve Demme

Points, Lines, Rays, and Line Segments

Geometry comes from the Greek word γεωμετρια for "earth measure." "Geo" means earth and "metry" means measure. To measure the earth, we need to break it into smaller, more manageable pieces.

Points

The smallest unit of measure is an imaginary piece called a ***point***. It has no measurable size, only position, or location. We can't measure its width or length, so it has no dimensions, or is zero-dimensional. To show something that is so small you can't really see it, we draw a dot. The dot is the "graph" of the point; it represents the point. We call it "point A" and label it with a capital, or uppercase, letter.

Figure 1   • A   point A

Lines

Using the point as the building block, consider a lot of connected points. A ***line*** is defined as an infinite (∞) number of connected points. It can be curved or straight. For our purposes, when we refer to a line, it will be straight unless mentioned otherwise. Two points that are contained in the same line are said to be ***collinear***. Since a line is as wide as a point, it has no width. But a line does have one dimension, which is length, so it is one-dimensional. A line is drawn, or "graphed," with arrows at both ends to show that it goes on indefinitely, or infinitely.

To label a line, use a lowercase letter or choose two points (represented with uppercase letters) in the line. Figure 2 we call "line *m*" and figure 3 "line QR," or $\overleftrightarrow{QR}$. In figure 3, the order of the points is not important. It could also be named "line RQ," or $\overleftrightarrow{RQ}$.

Figure 2

Figure 3

Some other figures which relate to the line are rays and line segments.

Rays

A *ray* is often referred to as one-half of a line. It has a specific starting point at one end, called the *endpoint*, or *origin*, and then proceeds infinitely in the other direction. Think of a ray as a flashlight or laser beam. Figure 4 is labeled as $\overrightarrow{BC}$ and read as "ray BC." When you label a ray, the order of the points is very important. The first letter is always the origin.

Figure 4

Line Segments

A *line segment* is a finite, or measurable, piece of a line. A line segment is not infinite like a line. It proceeds from one endpoint to another endpoint. It has a specific length. Figure 5 is labeled as $\overline{LH}$ and read as "line segment LH," or "segment LH."

Figure 5

Symbols

When speaking of two shapes that are exactly the same, we say they are *congruent*. The symbol for congruent is ≅. It comes from putting ~ and = together. The symbol ~ by itself means similar, or the same shape but not exactly the same size. Two squares have exactly the same shape but may have different measurements. Consider a house and a picture of a house. They have the same shape but are not the exact size, so they are *similar*.

The equals sign (=) means exactly the same length, or equal, and is used if two line segments are the same measurable length. Putting the symbols together means exactly the same shape and size, which gives us congruent (≅). Use the equals sign for measurable objects with the same measure. Choose the congruent sign for objects that have the same shape and size.

AB means the distance between A and B. If there is no symbol over the AB as in a ray, line, or line segment, then AB is the distance between the two points.

Planes and Sets

A *plane* is an infinite number of connected lines lying in the same flat surface. A plane has length and width, so it has two dimensions. A plane can be curved, as in a rolled up piece of paper, but for our purposes it will be flat. We graph a plane by drawing a parallelogram and labeling it with a lowercase letter. In figure 1, the plane is referred to as "plane *b*."

Figure 1

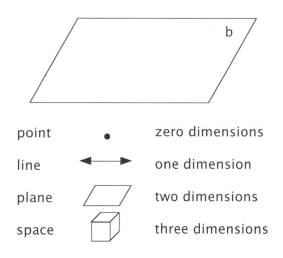

point	•	zero dimensions
line	←→	one dimension
plane	▱	two dimensions
space	⬚	three dimensions

Two lines that intersect are contained in the same plane. Two lines that lie in the same plane are said to be *coplanar*. The intersection of two planes is a line.

Most of our attention will refer to flat, two-dimensional shapes that lie in a plane. This is called *plane geometry*. Three-dimensional geometry, with length, width, and height (or depth), pertaining to space and solids, is called space geometry, or *solid geometry*. Solid geometry applies to volume, as of a cube, cylinder, pyramid, cone, or sphere, all of which will be covered later.

Set Symbolism

In the beginning of our study, we are being introduced to new concepts and shapes. We are also learning new symbolism and vocabulary. In lesson 1, we learned similar (~) and congruent (≅). In this lesson, there are five new symbols. They are ∩, ∪, ∅, ⊂, and { }.

Intersection - ∩ - This is where two or more things or groups meet or overlap.

Union - ∪ - This is where two or more things or groups are combined.

Empty set, or null set - ∅ - This means there is no possible answer.

Subset - ⊂ - This means one set is a subset of another.

Set - { } - These are the symbols that represent a set.

These symbols are usually introduced in a discussion of *sets*. A set is a collection of things. Each of the things is called an element, or *member* of the set. Normally in math, capital letters are used to represent each set. As an example, the set D, for Demme, will represent all the members of my immediate family. Set D, or just D = {Steve, Sandra, Isaac, Ethan, Joseph, and John}. Remember that braces ({ })are used as the symbol for a set. This is a finite set consisting of six members. If I were to show the infinite set of even numbers, it would be E = {2, 4, 6. . .} with the dots meaning that it goes on infinitely.

Another symbol is ⊂ , which represents a subset. In the first set D, Steve and Sandra are parents (set P) and the four boys are the children (set C). We could say the parents (P) are a subset of, or "are contained in," D. This is written as P ⊂ D. If we are speaking of the children, it would be C ⊂ D, which is the children included in the Demmes. We also might describe the Demmes in terms of five males and one female. If the question is asked, "How many female children?" and there aren't any, then the answer can be written as { } (empty set) or ∅ (empty or null set).

Intersections and Unions

Let's use geometry to illustrate *intersection* ($\cap$) and *union* ($\cup$) in the following examples. Refer to figure 2.

Figure 2

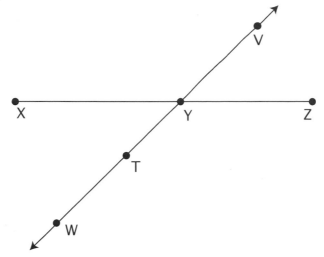

Example 1

The union of line segment $\overline{XY}$ and line segment $\overline{YZ}$ is line segment $\overline{XZ}$. Using the symbols, $\overline{XY} \cup \overline{YZ} = \overline{XZ}$. Notice when referring to union, the two elements are combined, or united.

Example 2

$\overrightarrow{TW} \cup \overrightarrow{TY} = \overleftrightarrow{WY}$. ($\overleftrightarrow{TY}$ and $\overleftrightarrow{TW}$ represent the same line as $\overleftrightarrow{WY}$.)
Two collinear rays that are united like this are a line.

Example 3

$\overrightarrow{YT} \cup \overrightarrow{TW} = \overrightarrow{YW}$
Do you see that the inclusion of $\overrightarrow{TW}$ on $\overrightarrow{YT}$ does not change $\overrightarrow{YT}$ since it is already going to infinity. Ray TW is already included in ray YT; joining them does not change the shape of ray YT.

Example 4

The intersection ($\cap$) of $\overline{XY}$ and $\overline{TY}$ is point Y. This intersection is what they share, or have in common. It is symbolized as $\overline{XY} \cap \overline{TY} = Y$.

Example 5

The intersection of $\overrightarrow{YW}$ and $\overrightarrow{TV}$ is $\overline{YT}$, written symbolically
as $\overrightarrow{YW} \cap \overrightarrow{TV} = \overline{YT}$.

Example 6

The intersection of $\overrightarrow{TW}$ and $\overline{XZ}$ is nothing, written symbolically
as $\overrightarrow{TW} \cap \overline{XZ} = \varnothing$ or { }. The answer is the empty, or null, set since the
ray and line segment don't intersect.

Example 7

Using subsets: $\overline{XY} \subset \overline{XZ}$ and $\overline{TY} \subset \overrightarrow{TY}$ since the first members are
contained in the second members.

LESSON 3

Angles

If two lines intersect, the opening or space between them is an *angle*. In figure 1 there are four angles shown by the arcs. The angles are named 1, 2, 3, and 4.

Figure 1

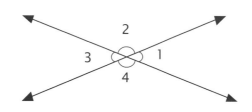

In figure 2 we are focusing on just one angle, which is made by drawing two rays with a common endpoint. This endpoint, or origin, is called the *vertex.* (The plural of vertex is "vertices.")

Figure 2

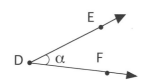

The rays are $\overrightarrow{DE}$ and $\overrightarrow{DF}$. The angle is labeled with either a number as in figure 1 or a lowercase Greek letter as in figure 2. We'll call this figure ∠α ("angle alpha"). Another way to identify this angle is by picking one point on each ray and the vertex, so: ∠EDF or ∠FDE. Notice that the point labeling the vertex is always in the middle.

Measuring Angles

At this point, it would be a good idea to practice reading angles, so you can easily identify them. It is also a good time to meet a protractor, which is used to measure angles. You probably already know how to use a ruler. *Rulers* measure length, but *protractors* measure angles.

Refer to figure 3, and measure the length of the line segments AB and BC, which are parts of the two rays forming ∠ABC.

Angles are measured in *degrees*. The wider the gap, or opening, the more degrees. The smaller the opening, the fewer degrees there are. The "measure of an angle" is represented by "m∠." To represent "the measure of angle one is thirty degrees," write "m∠1 = 30°."

Using a protractor, find m∠ABC in figure 3. Then find m∠DEF and m∠GHJ in figures 4 and 5.

Figure 3

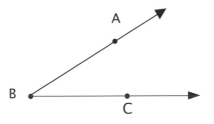

Figure 4

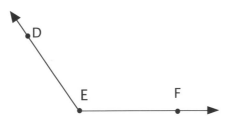

Figure 5

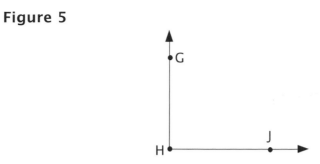

Drawing Angles

Now practice drawing angles with your ruler or straightedge and your protractor. Draw ∠QRS with measure 50°.

1. Draw a straight line with one endpoint (a ray).

2. Then place your protractor on the line with the cross hairs on the endpoint, and the line going through 0° and 180°.

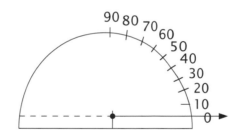

3. Beginning with 0° on the ray, count up to 50° and put a mark on your paper by 50°.

4. Remove the protractor, and using your straightedge, connect the vertex with the mark at 50°.

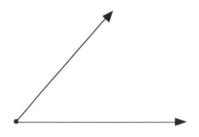

5. Label the angle with points.

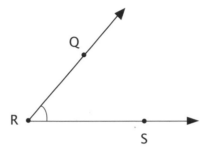

We placed a small arc to show that the angle we want to refer to is inside the rays and not outside them.

6. Then we write the angle with the measure: "m∠QRS = 50°," which symbolizes "the measure of angle QRS is 50 degrees."

Types of Angles

Right Angle

A **right angle** has a measure of 90°. It is the angle used most often. When two rays or line segments that form an angle have a measure of 90°, they form a right angle, or square corner, as in figure 1. Usually a box at the vertex is used to represent the right angle. Notice that it doesn't matter where the angle is, but only that its measure is 90°.

Figure 1

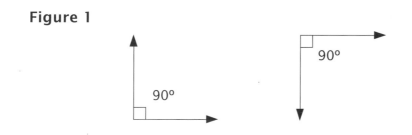

Straight Angle

A **straight** *angle* is a lesser known angle, which is difficult to think of as an angle. It has a measure of 180°.

Figure 2

You sometimes hear of a car that skidded on ice and did a "one-eighty," meaning the car was going in one direction and then spun around so it was pointing in the opposite direction. This expression comes from the fact that the car spun 180°. Figure 3 is a top view of this skid, and figure 4 is the side view.

Figure 3

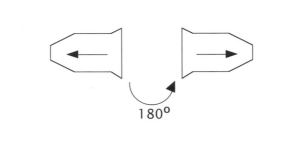

180°

Figure 4

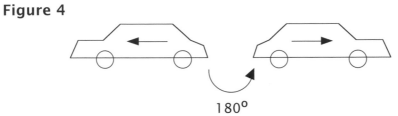

180°

360° Angle

In basketball, there are some players that run, jump, turn completely around, and land in the same direction as when they started. This is called a "360" since the players turned completely around in the air, or **360°**, and are still facing the basket when they land.

Figure 5

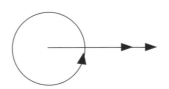

All the way around is 360°.

Acute Angle

An *acute angle* is more than 0° but less than 90°. Most of the angles you see are acute angles. Since they are small angles, it helps me to remember the name by thinking of "cute." Figure 6 shows two acute angles.

Figure 6

$$0° < \text{acute} < 90°$$

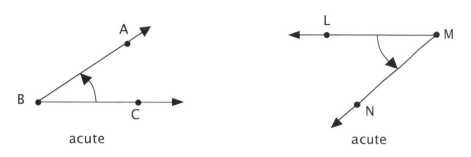

acute   acute

Obtuse Angle

An *obtuse angle* is larger than 90° and less than 180°. See figure 7.

Figure 7

$$90° < \text{obtuse} < 180°$$

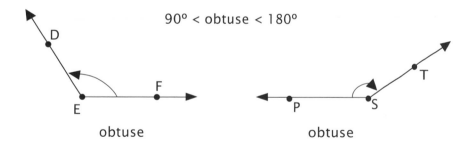

obtuse   obtuse

Reflex Angle

A *reflex angle* is larger than 180° and less than 360°.

180° < reflex < 360°

Figure 8

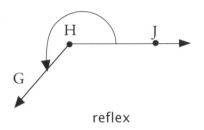

reflex

Figure 9

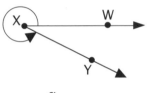

reflex

Measure a Reflex Angle

To measure a reflex angle, you have two options. The first is to measure either the obtuse angle, as in figure 8, or the acute angle as in figure 9, and subtract from 360°. The second option, as shown in figure 10, is to extend one ray to point F. Then measure ∠FHG and add it to 180° (the measure of ∠JHF).

Figure 10

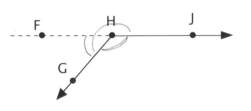

Parallel and Perpendicular Lines
with Midpoints and Bisectors

Parallel lines are defined as two lines in the same plane that never intersect. With set symbolism, we could say of figure 1: $\overleftrightarrow{AB} \cap \overleftrightarrow{CD} = \varnothing$.

Figure 1

Although not exact, some examples of parallel lines are railroad tracks and the opposite lanes of an interstate highway with a median. If you use lined paper, the lines should be parallel. You hope, when you build a house, that the planes representing the first floor and the ceiling are parallel! The symbol for parallel is ||. We think of this symbol as representative of the two *l*s in the word parallel. In describing figure 1, we say $\overleftrightarrow{AB} \parallel \overleftrightarrow{CD}$.

Perpendicular Lines

Two lines or rays or line segments that intersect and form a right angle are called *perpendicular lines*. When we have perpendicular lines, we draw a little square where the lines meet to indicate they form a right angle (as in lesson 4). The symbol for perpendicular is ⊥.

Figure 2

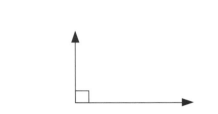

Examples of perpendicular lines may be found in many places. A few obvious ones are the intersection of two roads in a town, where the walls in a house meet, or telephone poles and their cross trees. Perpendicular lines are very important in construction. Contractors call this "making things square." Walls, doors, windows, and corners all have to be square!

Midpoint

Recall a line segment, which has specific length and two endpoints. Of all the points, the one in the very middle is the ***midpoint.*** It divides the line segment into two line segments that have the same length, or are congruent. In figure 3, note the relationship between $\overline{AH}$ and $\overline{GH}$.

Figure 3

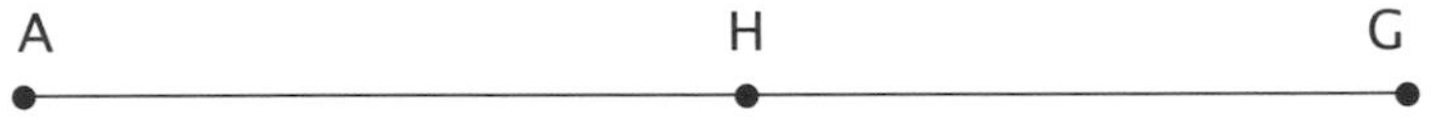

A   H   G

If H is the midpoint, we can conclude that $\overline{AH} \cong \overline{HG}$ and AH = HG. The distance between A and H is the same as the distance between H and G.

Bisector

If an angle is cut exactly in half, we say that it has been **bisected**, and the line or ray which cuts an angle in half is called a **bisector**. In figure 4, ∠YHM is bisected by HB. If the measure of ∠YHM is 64°, then ∠YHB and ∠MHB would each measure 32°. So we can say m∠YHB = m∠MHB and ∠YHB ≅ ∠MHB

Figure 4

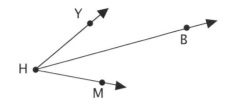

Constructing a Perpendicular Bisector

Now let's use a compass to find the midpoint and bisector of a line segment.

Example 1

1. Draw a line segment two inches long and label the endpoints E and W.

2. Placing the sharp end of the compass at point E, draw an arc that is over half the distance from E to W.

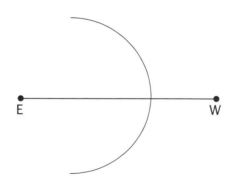

3. Without changing the spread of your compass, move the sharp end to W and draw another arc.

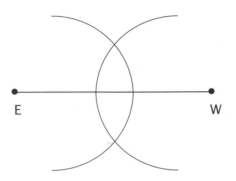

4. Notice the points where the arcs meet above and below $\overline{EW}$. We'll label the points N and S. When we draw a line through N and S, it intersects $\overline{EW}$ at point C. Point C is the midpoint of $\overline{EW}$; therefore $\overline{EC} \cong \overline{WC}$ and EC = WC. Notice also that $\overset{\leftrightarrow}{NS} \perp \overset{\leftrightarrow}{EW}$.

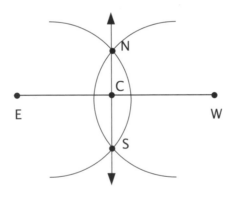

Since $\overset{\leftrightarrow}{NS}$ bisects $\overline{EW}$ and is also perpendicular to it, $\overset{\leftrightarrow}{NS}$ is referred to as a *perpendicular bisector.*

Bisecting an Angle

Now we'll construct the bisector of an angle using only a straight edge and a compass.

Example 2

1. Using a straight edge or ruler, draw two rays to form an angle.

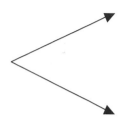

2. Set your compass so the distance between the point and the pencil is greater than half the length of either ray. With the compass point on the vertex of the angle, draw an arc that will intersect both rays.

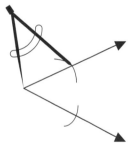

3. Putting the point of the compass where the arcs intersect the rays, make two more arcs that intersect inside the angle away from the vertex. Connect that point of intersection with the vertex of the angle by drawing a line or ray. This is the line or ray that bisects the angle, and it is called the bisector.

Supplementary and Complementary Angles

Greek Letters

Figure 1

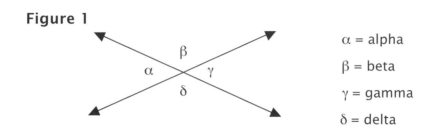

α = alpha

β = beta

γ = gamma

δ = delta

Adjacent Angles

Angles that share a common side and have the same origin are called ***adjacent angles.*** They are side by side. In figure 1, α is adjacent to both β and δ. It is not adjacent to γ. In figure 1, there are four pairs of adjacent angles: α and β, β and γ, γ and δ, δ and α.

In figure 2, we added points so we can name the rays that form the angles. The common side shared by adjacent angles α and β is $\overrightarrow{VQ}$.

Figure 2

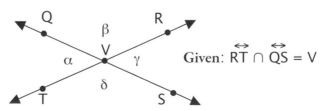

Given: $\overleftrightarrow{RT} \cap \overleftrightarrow{QS} = V$

Vertical Angles

Notice that ∠γ is opposite ∠α. Angles that share a common origin and are opposite each other are called *vertical angles*. They have the same measure and are congruent. ∠β and ∠δ are also vertical angles.

Figure 2 **(from previous page)**

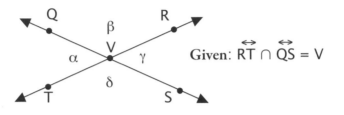

Given: $\overleftrightarrow{RT} \cap \overleftrightarrow{QS} = V$

If m∠β is 115°, then m ∠δ is also 115°. If this is true, then do we have enough information to find m ∠α? We know from the information given in figure 2 that $\overleftrightarrow{RT}$ and $\overleftrightarrow{QS}$ are lines. Therefore, ∠RVT is a straight angle and has a measure of 180°. If ∠RVQ (∠β) is 115°, then ∠QVT (∠α) must be 180° - 115°, or 65°. Since ∠RVS (∠γ) is a vertical angle to ∠QVT, then it is also 65°.

Supplementary Angles

Two angles such as ∠α and ∠β in figure 2, whose measures add up to 180°, or that make a straight angle (straight line), are said to be *supplementary*. In figure 2, the angles were adjacent to each other, but they don't have to be adjacent to be classified as supplementary angles.

Figure 3

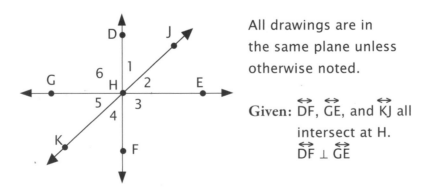

All drawings are in the same plane unless otherwise noted.

Given: $\overleftrightarrow{DF}$, $\overleftrightarrow{GE}$, and $\overleftrightarrow{KJ}$ all intersect at H.
$\overleftrightarrow{DF} \perp \overleftrightarrow{GE}$

Complementary Angles

We can observe many relationships in figure 3. Angle 1 is adjacent to both ∠6 and ∠2. Angle 3 and ∠6 are vertical angles, as are ∠1 and ∠4. Angle 6 and ∠3 are also right angles since $\overleftrightarrow{DF} \perp \overleftrightarrow{GE}$. The new concept here is the relationship between ∠DHE and ∠GHF. Both of these are right angles because the lines are perpendicular; therefore their measures are each 90°. Then m∠1 + m∠2 = 90°, and m∠4 + m∠5 = 90°. Two angles whose measures add up to 90° are called ***complementary angles***. Notice that from what we know about vertical angles, ∠1 and ∠5 are also complementary. Let's use some real measures to verify our conclusions.

Figure 4   (a simplified figure 3)

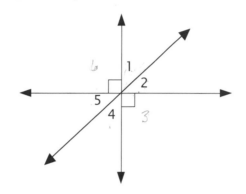

In figure 4, let's assume that m∠1 = 47°. Then m∠2 must be 43°, since m∠1 and m∠2 add up to 90°. If m∠1 = 47°, then m∠4 must also be 47°, since ∠1 and ∠4 are vertical angles. Also, m∠5 must be 43°. So ∠1 and ∠5 are complementary, as are ∠2 and ∠4. Remember that supplementary and complementary angles do not have to be adjacent to qualify.

It helps me to not get supplementary and complementary angles mixed up if I think of the *s* in straight and the *s* in supplementary. The *c* in complementary may be like the *c* in corner.

Transversals
With Interior and Exterior Angles

A *transversal* is a straight line that intersects two or more parallel lines. Two parallel lines that lie in the same plane and are cut by a transversal produce several interesting angles.

Figure 1

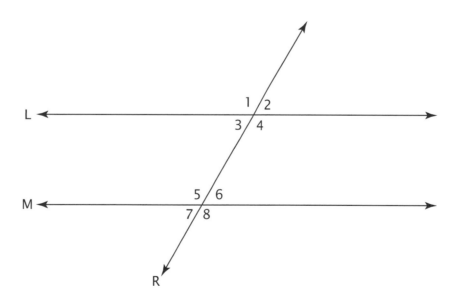

Given: Lines L and M are ||.

Line R intersects each of them and is a transversal.

Interior and Exterior Angles

Angles 3, 4, 5, and 6 are inside the parallel lines and are called *interior* angles. Angles 1, 2, 7, and 8 are outside the parallel lines and are called *exterior* angles. See figure 2.

Figure 2

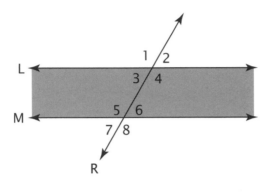

Corresponding Angles

Notice the cluster of four angles around the intersection of lines L and R in figure 2. Also notice the cluster around the intersection of lines M and R. Which angle in the first group *corresponds* to ∠7 in the second group? The answer is ∠3, since both angles are on the lower left-hand side. So ∠1 corresponds to ∠5, ∠2 corresponds to ∠6, and ∠4 corresponds to ∠8.

If we know the measure of ∠1, we can use what we learned in lesson 6 to find the measures of the other seven angles. Look at figure 3. If m∠1 is 120°, then m∠2 is 60° because they are supplementary, and m∠4 is 120° because it is opposite ∠1 and is a vertical angle. Since m∠2 = 60°, then m∠3 = 60° since they are vertical angles. This much we already know. What this lesson is teaching us is the relationship between corresponding angles. I picture the relationship as if line M was originally directly atop line L, and angles 5, 6, 7, and 8 were covering angles 1, 2, 3, and 4. Since these two lines are oriented in exactly the same direction, they are identical. Since one line moved down, it is still identical, has the same angles, and is an exact image of the other. See figure 3.

Figure 3

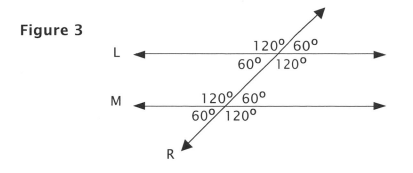

Alternate Angles

Looking at figure 4, notice that ∠3 and ∠6 are interior angles, and are on opposite sides of the transversal. They are called *alternate interior* angles. The other pair of alternate interior angles shown are ∠4 and ∠5.

Which do you think are the two pairs of *alternate exterior* angles? One pair is ∠1 and ∠8, and the other pair is ∠2 and ∠7.

Figure 4

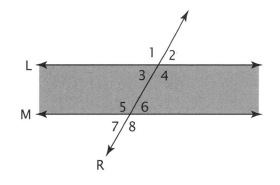

A Postulate

So one of our first observations, or **postulates**, is that when two parallel lines are cut by a transversal, corresponding angles are congruent. Building on this postulate, it follows that alternate interior angles are congruent, and alternate exterior angles are also congruent. Look at the measurements in figure 5 verify this.

Figure 5

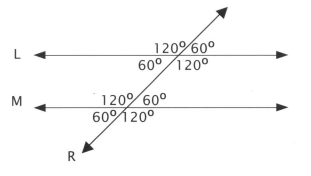

A Converse of a Postulate

Postulates are important, but so are their **converses**, or opposites.

Here is the converse of our first postulate: if two parallel lines cut by a transversal produce congruent corresponding angles, then congruent corresponding angles produce two parallel lines. In figure 4, if it were given that ∠3 and ∠7 are congruent, then we would know that lines L and M are parallel.

Perimeter; Interior Angles
Rectangle, Triangle, Parallelogram, and Trapezoid

This should be a review lesson for most students, but while going over familiar ground we will also make sure we know the names of and are able to identify these common quadrilaterals: square, rectangle, parallelogram, rhombus, and trapezoid.

Perimeter

Perimeter comes from the two roots "peri" and "meter." "Meter" means measure and "peri" means around. So perimeter is the measure of distance around a shape, or figure. Perimeter is especially applicable when measuring for a fence or for baseboard in a room. Because it is the distance, or line around, the answer is always given as linear units or linear feet, etc. Another tip to keep from confusing perimeter and area is to think of the word RIM within perRIMeter.

Shapes

Now let's define our shapes. A *quadrilateral* has four sides, as the name reveals: "quad" means four and "lateral" means sides. A *rectangle* is a quadrilateral with four right angles. A *parallelogram* is a quadrilateral with two pairs of parallel sides. A *rhombus* is a quadrilateral with four congruent sides. A *trapezoid* is a quadrilateral with only one pair of parallel sides. A *square* may be defined several ways. It could be a rectangle (four right angles) with four congruent sides, or it could be a rhombus (four congruent sides) with four right angles. It could also be described as a quadrilateral with four congruent sides and four right angles. All of these definitions are correct.

We know that the opposite sides of a parallelogram and of a rectangle have the same lengths. We will prove this formally in lesson 25.

Here are some examples of finding the perimeter. Perimeter is found by adding up the lengths of all the sides, so ignore any other measurements that are given.

Example 1

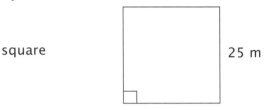

square

P = 25 + 25 + 25 + 25 = 100 m, or P = 4(25) = 100 m

Example 2

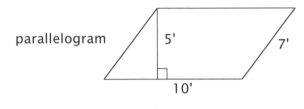

parallelogram

P = 10 + 7 + 10 + 7 = 34 ft

Example 3

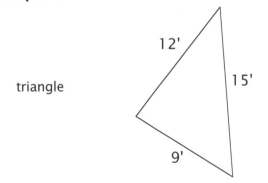

triangle

P = 12 + 15 + 9 = 36 ft

Example 4

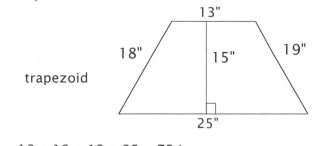

trapezoid

P = 18 + 19 + 13 + 25 = 75 in

Interior Angles

Notice that if a rectangle has four right angles, then the ***sum of the interior angles*** is 360°. This is true for all of these quadrilaterals. We deduce this by dividing the quadrilaterals into two triangles by drawing a straight line connecting the opposite vertices.

Figure 1

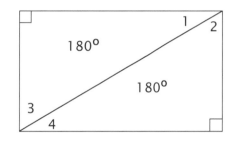

The sum of the angles of a triangle is always 180°. If m∠1 is 40°, then m∠2 must be 50°. Because ∠1 and ∠4 are alternate interior angles, m∠4 is 40° and m∠3, therefore, is 50°.

Figure 2

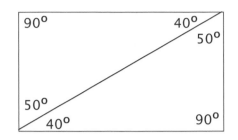

Adding the angles all up, we see 180° in each triangle and 360° in the quadri-lateral. This holds true for all of the following quadrilaterals even if they do not have right angles, because all quadrilaterals can be divided into two triangles with 180° each.

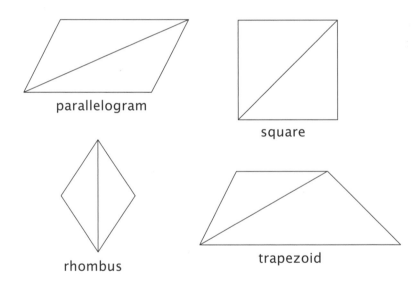

parallelogram

square

rhombus

trapezoid

LESSON 9

Area
Rectangle, Triangle, Parallelogram, and Trapezoid

Area of a Rectangle

The *area* of a rectangle is found by using the over and the up, or the *base* times the *height*. To remember that the height is always perpendicular, look at the small letter "h" in the word "height." When showing perpendicular lines, we put a square where the lines intersect.

Figure 1

When we do this, we create the letter "h." So the height is always perpendicular to the base. This is very important when finding the area of a triangle, a parallelogram, or a trapezoid. The height is not one of the sloping sides; it is always the perpendicular dimension.

Example 1

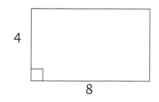

Area = bh = 8 x 4 = 32 square units (units^2)

Area of a Parallelogram or Rhombus

To find the area of a parallelogram or rhombus, we use the same formula as we did for finding the area of a rectangle: over times up, or base times height. To understand the formula, look at the following picture. The dotted lines show where we cut off the piece on the left and slid it to the right to make a rectangle. We know the area of a rectangle is base times height.

Figure 2

When finding the area of a parallelogram or trapezoid, remember that the height is always perpendicular to the base. Consider the following example:

Example 2

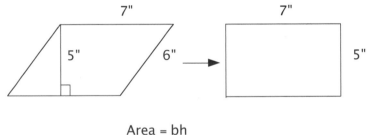

Area = bh

= 7 × 5 = 35 square inches $\left(in^2\right)$

Area is base times height.

Area of a Triangle

The area of a triangle is found by multiplying the base times the height times the fraction one-half.

Example 3

To find the area of the white triangle, picture it as half of a rectangle. Find the area of the rectangle by multiplying the base times the height. The triangle is half of this, so b x h x 1/2 gives us the area of a triangle.

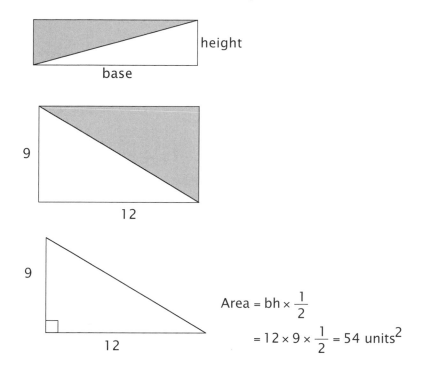

$$\text{Area} = bh \times \frac{1}{2}$$
$$= 12 \times 9 \times \frac{1}{2} = 54 \text{ units}^2$$

Area of a Square

The area of a square is also found by multiplying the over times the up, or the base times the height. Since the base and the height are the same length in a square, the formula for the area of a square can also be expressed as side times side, or side squared.

Example 4

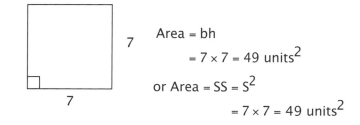

$$\text{Area} = bh$$
$$= 7 \times 7 = 49 \text{ units}^2$$
$$\text{or Area} = SS = S^2$$
$$= 7 \times 7 = 49 \text{ units}^2$$

Notice that when finding area, the answer is always in square units, such as square inches (in^2), square feet (ft^2), etc.

Area of a Trapezoid

To find the area of a trapezoid, we use the same formula as for a rectangle—with a slight change. Look at figure 3 to understand the formula. We cut two corner pieces, one from the left and one from the right. Then pivot them up as if they were on hinges to create a rectangle.

Figure 3

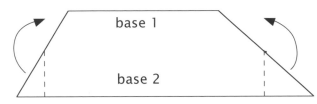

So the base is not the top flat piece (read as "base one") or the long bottom piece (read as "base two"), but the average base, or the one in the middle. The formula is the average base times the height. The formula for finding the area of a trapezoid is traditionally expressed as: $\frac{B_1 + B_2}{2} \times h$

Figure 4

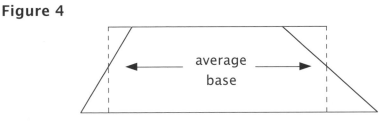

Adding the two bases and dividing by two is the formula for finding the average base.

Example 5

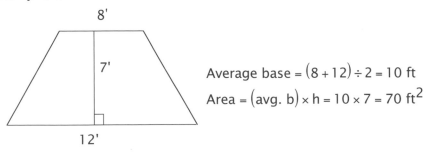

Average base = $(8 + 12) \div 2 = 10$ ft

Area = (avg. b) $\times$ h = $10 \times 7 = 70$ ft^2

Constructing and Identifying Triangles

Constructing a Triangle

A triangle consists of three angles and three sides. Construct a triangle that is large enough so that you can measure each angle with your protractor. Measure each angle and each side. Do you notice a relationship between an angle and the side opposite the angle? The smaller the angle is, the smaller the side opposite that angle. The larger the angle is, the larger the side opposite that angle.

Identifying Triangles

Triangles may be described in terms of their sides or their angles. A triangle with all sides congruent and all angles congruent (regular polygon) is identified as *equilateral* (equal sides) or *equiangular* (equal angles). Usually this triangle is called equilateral. Since the interior angles add up to 180° and each angle has the same measure, the measure of each angle is 60°.

Figure 1

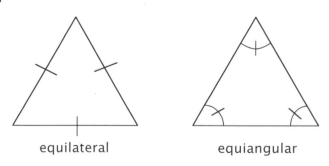

equilateral   equiangular

We use slash marks and arcs to show parts that are congruent.

If two sides and two angles are congruent, it is an *isosceles triangle.*

Figure 2

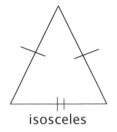

isosceles

A triangle with no sides congruent is called a *scalene triangle.*

Figure 3

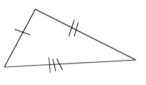

scalene

A triangle with all three angles greater than 0°, and each less than 90°, is an *acute triangle.* This can be expressed as: each angle > 0° and < 90°, or 0° < each angle < 90°.

Figure 4

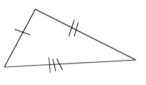

acute triangle

A triangle with one angle greater than 90° is an *obtuse triangle.* This can be expressed as: one angle > 90° and <180°, or 90° < one angle < 180°.

Figure 5

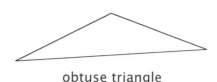

obtuse triangle

A triangle with one angle equal to 90° is a *right triangle.* This can be expressed as: one angle = 90°.

Figure 6

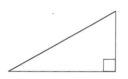

right triangle

Could there be an acute triangle that is also equilateral? Yes, because if the triangle is equilateral, then all three angles have a measure of 60°, which means they are all acute as well. Could there be an isosceles right triangle? Yes, if the legs are congruent, the right triangle will also be isosceles.

Limitations

We know that there are limitations in the measures of the angles in a triangle. If one angle is 80°, and another is 70°, then the third angle must be 30°. There are also limitations on the lengths of the sides of a triangle. I'd like you to figure this one out yourself. Construct a triangle using three of the unit bars: the orange two bar, the light blue five bar, and the tan seven bar. Can you do it? You can see that you will need at least the pink three bar in place of the two bar.

Figure 7

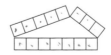

With sides of 5 and 7, the small side must be > 2.

If you have just the five bar and the seven bar, you can also see that the third bar can't be any longer than twelve, and using whole numbers, must be at the most eleven units long.

Figure 8

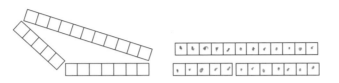

With sides of 5 and 7, the third side must be < 12.

Try two five bars with the nine bar. Can you do it now? Can you develop a principle based on these observations? The two smaller sides together must be longer than the third side.

We can represent this observation as follows: if side A ≤ side B ≤ side C (so A is smaller or equal to side B, and C is the longest), then A + B > C (the two smaller sides added up must be longer than C). Consider the lengths of the sides of another triangle: 6 in–7 in–11 in. Could these be the dimensions of a triangle? Yes, because the sum of the two smaller sides, 6 + 7, is longer than 11, or 6 + 7 > 11. What about a triangle with sides of 6 cm–3 cm–11 cm? No, these could not be the dimensions of a triangle, because 3 + 6 < 11.

Figure 9

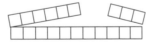

With the long side 11, the combined length
of the other two sides can't be < 11.

Regular Polygons

The word *polygon* comes from two Greek words: "polus" meaning many and "gonia" meaning angles. It literally means many angles. It is also defined in general terms as a closed curve. There are two kinds of closed curves: *concave* and *convex*. Concave shapes have "caves." In this book, we will be studying convex polygons.

Figure 1

cave

concave

convex

Regular Polygons

   A polygon with all sides congruent and all angles congruent is a *regular polygon*. In this course, whenever a polygon is considered, it will be a regular polygon.

Interior Angles

   To find the measure of each angle in a regular polygon, find the total number of degrees of the interior angles and divide by the number of angles. We know the number of degrees in a triangle is 180°. If it is an equilateral, equiangular triangle (regular polygon with three sides), each angle would be 180° divided by 3, or 60°.

Figure 2

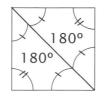

180°

180°

A square can be made into two triangles by drawing one diagonal (a line connecting two vertices). The measure of each triangle is 180°. Therefore, the total number of degrees in a square is 360°. Of course, we know this because we know a square has four right angles, or 4 x 90°, or 360°.

Notice that we wouldn't want another diagonal that would create four triangles (4 x 180° = 720°), because we want only the measure of the angles on the square, and not all of those inside as well.

Figure 3

The angles without slashes are not on the outside of the polygon and should not be counted.

Kinds of Polygons

Here are some of the more common convex polygons: *quadrilateral, pentagon, hexagon, octagon, decagon,* and *dodecagon.* A decagon has 10 sides and a dodecagon has 12 sides. Under the heading of quadrilateral, we find *rectangle, square, parallelogram, rhombus,* and *trapezoid.* The *square* is the quadrilateral that has all four sides congruent and all four angles congruent.

Figure 4

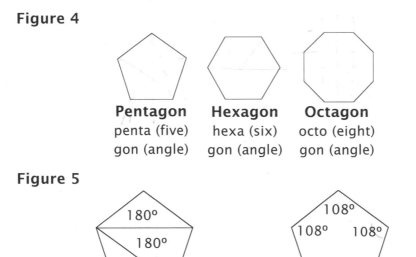

Pentagon
penta (five)
gon (angle)

Hexagon
hexa (six)
gon (angle)

Octagon
octo (eight)
gon (angle)

Figure 5

In a pentagon, we can draw two diagonals to make three triangles. So we have 3 x 180°, or 540°. Since there are five interior angles, 540° ÷ 5 = 108° per angle.

Figure 6

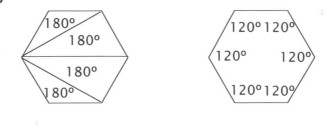

In a hexagon, we can draw three diagonals to make the four triangles, and 4 x 180° or 720°. Since there are six interior angles, 720° ÷ 6 = 120° per angle.

Take notice of the relationship between the number of sides and the number of triangles:

Shape	Sides	Triangles
triangle	3	1
square	4	2
pentagon	5	3
hexagon	6	4
octagon	8	6
decagon	10	8
dodecagon	12	10

You can see that the number of sides, minus two, equals the number of triangles. Multiplying the number of triangles by 180° gives us the total degrees in the polygon. So our formula is $(N - 2) \times 180°$, where N = Number of sides. We would expect the total number of degrees of an octagon to be $(8 - 2) \times 180°$, or 1,080°. Each individual angle in an octagon would be 1,080° divided by 8, or 135°.

$$(N - 2) \times 180° = \text{Total Degrees}$$

Exterior Angles

Up till now, we've been calculating the measure of the interior angles only. But each of these shapes also has *exterior angles*, which can be drawn by extending each of the line segments. Since the interior angles of an equilateral triangle are each 60°, and the exterior angles are supplementary, then the three exterior angles are each 120°; therefore the three exterior angles add up to 360° (3 x 120°).

Figure 7

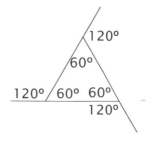

Since each of the interior angles of a pentagon is 108°, then each of the supplementary exterior angles is 72°. The five exterior angles add up to 360° (5 x 72°).

Figure 8

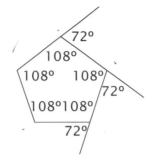

If each of the interior angles of a square is 90°, then each of the supplementary exterior angles is 90°. So the four exterior angles add up to 360° (4 x 90°).

Figure 9

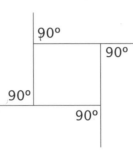

Exterior angles always add up to 360°. Knowing this, you can think conversely to find the measure of each interior angle. In an octagon, the eight exterior angles would add up to 360°. Each exterior angle would be 360° ÷ 8. or 45°. Each interior angle would be supplementary to an exterior angle with measure 45°.

INT + EXT = 180°, and INT + 45° = 180, so the measure of each interior angle is 135°.

Figure 10

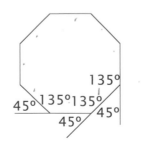

Geometry of a Circle, Sphere, and Ellipse
Inscribed and Circumscribed Figures

Circle

Picture a nail, a piece of string, and a pencil. If you were to tie the string to the nail at one end and the pencil to the other, and then move the pencil around the nail with the string stretched taut, you would be drawing a circle. The length of the string would be the *radius*. The nail would be at the *center*. We might define a *circle* as a set of points (pencil marks) an equal distance (length of string) from the center point (location of the nail). See figure 1.

Figure 1

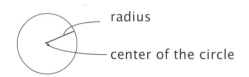

radius

center of the circle

Measures of a Circle

There is a difference between the set of points that are on the circle and the space inside the circle. To keep a clear distinction, we'll call the set of points drawn by the pencil the *circumference* of the circle, and the interior of this closed curve the *area* of the circle. The circumference and the area together we'll refer to as "the circle."

Figure 2

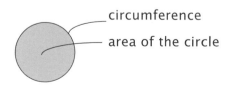

circumference

area of the circle

Line Segments of a Circle

A **chord** is a line segment drawn between two points on the circumference of the circle. If the chord goes through the center of the circle, it is called the **diameter**. The diameter is the longest possible chord in a circle. The **radius**, which is the distance from the center, is one half the length of the diameter.

Figure 3

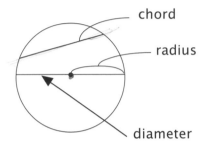

Lines of a Circle

A line that intersects the circumference of the circle at exactly one point is called a **tangent**. From this point of intersection to the center of the circle is the radius. At this point of intersection, the radius is perpendicular to the tangent. A line that intersects a circle in two points is called a *secant*. A secant is similar to a chord except a chord is a line segment, whereas a secant is a line. As such, a secant contains a chord.

Figure 4

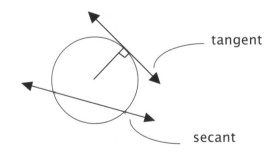

Pieces of a Circle

A circle has 360°. Think of the circle as a pie. If you wanted a piece of the pie, you'd measure it by degrees. Your chunk of pie would be a chunk of the area of the circle that is called a *sector*. A 90° sector looks like this:

Figure 5

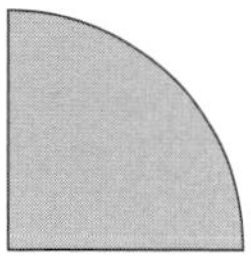

Sectors are pieces of the area.

Pieces of the circumference are called *arcs*. Here is a 90° arc:

Figure 6

Figure 6

Sphere

A three-dimensional circle is a *sphere* (like a ball).

Figure 7

Ellipse

We think of the orbits of the planets as circles; actually, they are irregular, or stretched, circles called *ellipses*. An oval race track would be a good example of an ellipse.

Figure 8

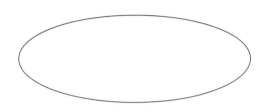

Central Angles

In figure 9, the circle is divided into two arcs, a *minor arc* and a *major arc*. If the measure of the minor arc is 60°, then the major arc would be 360° − 60°, or 300°, since the total number of degrees in a circle is 360°. In figure 10, point B is the center of the circle, and we can say m∠ABC = m$\overset{\frown}{AC}$. The curved line above $\overset{\frown}{AC}$ is the symbol for arc.

Figure 9

Figure 10

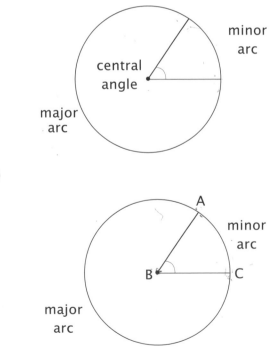

Inscribed and Circumscribed Figures

In figure 11, polygon PLYG is inside the circle, and we say it is *inscribed* inside the circle. To be inscribed, all the points, or vertices, of the polygon must lie on the circle. The circle, on the other hand, is around the polygon, and we say the circle is *circumscribed* around the polygon. In figure 12, the circle is inscribed inside the square, while the square is circumscribed around the circle.

Figure 11

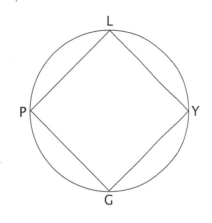

Figure 12

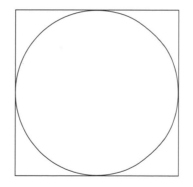

Inscribed Angles

In figure 13, triangle DEF is equilateral and equiangular. The measure of ∠DFE is 60°. Notice that the measure of arc DE is one-third of the circle, or 1/3 of 360°, which is 120°. What we observe is that the measure of an *inscribed angle* (∠DFE) is one-half the measure of the *intercepted arc* (DE).

In figure 14, the measure of the inscribed ∠QRS is 48°. The measure of the intercepted arc QS is two times 48°, or 96°.

Figure 13

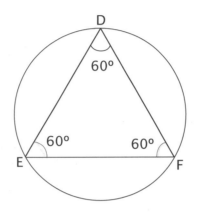

Figure 14

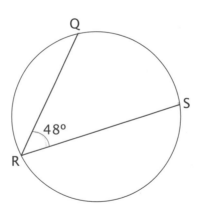

Historical Note

Archimedes lived from 287-212 BC. He was the first mathematician on record to estimate the value of π (taught in the next lesson) rigorously. He realized that its magnitude can be outer and inner polygons' respective perimeters. By using the equivalent of 96-sided polygons, he proved that 3 + 10/71 < π < 3 + 1/7. The average of these values is about 3.14185. (wikipedia)

LESSON 13

Area of a Circle and an Ellipse
Circumference of a Circle; Latitude and Longitude

π (sometimes written pi) is a mathematical constant whose value is the ratio of circumference to the diameter. π is an irrational number whose value cannot be expressed exactly as a fraction, and thus its decimal representation never ends or repeats. π is the first letter of the Greek word for perimeter "περμετροσ," and was introduced by William Jones in 1707.

Area

The formula for the area of a circle is πr^2. To help us remember the formula and understand where it originates, look at the picture of the circle inside the square (figure 1). Whenever finding area, remember the made-up word "SQUAREA." It reminds you that area is always given in *square* units: SQUARE + AREA = SQUAREA.

The area of the circle is a little more than the area of three of the squares with sides of length r (for radius) and area r^2. The value of π is a little more than three, or approximately 3.14.

Figure 1

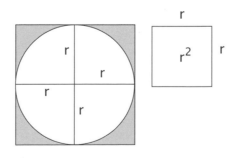

Example 1

The area of this circle is (3.14) (5")(5") = 78.5 in^2

            π r^2

Example 2

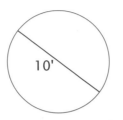

In this circle the diameter is given, so we first cut it in half to get a radius of 5', then square the 5 and multiply by 3.14 to get 78.5 ft^2.

$$(3.14)(5')(5') = 78.5 \text{ ft}^2$$

Circumference

The perimeter of a circle is called the ***circumference***. To find the circumference of a circle, multiply π by the diameter (πd), or multiply π by the radius times two (2πr), which is the same thing.

The value of π is approximately 22/7, or 3.1415927. . . . In our problems, we will round π to the hundredths place and use 3.14. Sometimes, depending on the problem, it may be more convenient to use 22/7.

Example 3

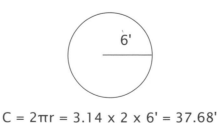

$$C = 2\pi r = 3.14 \times 2 \times 6' = 37.68'$$

Here we are given the radius, so we double it to get the diameter.

Example 4

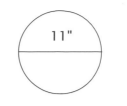

$C = \pi d = 3.14 \times 11" = 34.54"$

Notice that the answers to both problems are in feet and inches and are not squared; this is linear measure, used for perimeter and circumference.

Latitude

Since circumference and area are review topics for many of the students, this lesson also teaches students to read latitude and longitude. *Latitude* is the measure given in degrees for the distance north and south from the equator. To help me distinguish this from longitude, I remember "FLATitude," as the circles that measure latitude are horizontal, or flat.

Figure 1

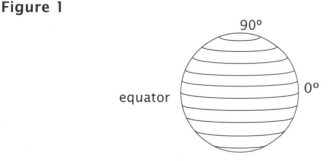

The equator is a circle that extends all around the earth in the middle at 0°. Beginning at the equator, lines of latitude measure from 0° to 90° north to the North Pole, and from 0° to 90° south to the South Pole. All the circles that represent latitude are parallel to each other, and they get smaller as they move towards the poles.

Longitude

The circles that represent *longitude* are LONG, and theoretically all are the same length.

Figure 2

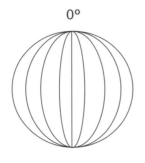

0°

Figure 3

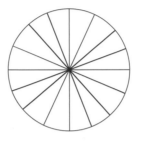

A view from the North Pole

Figure 4

Putting latitude and longitude together.

The circles that represent longitude go all around the globe and measure the distance from east to west. They are also measured in degrees and begin at 0° in Greenwich, England and extend 180° east and 180° west. The line of longitude that goes through Greenwich is called the prime meridian. For more detailed locations, each degree in broken down into 60 minutes (60') and each minute into 60 seconds (60"). New Orleans is almost exactly 30°N (starting at the equator) and 90°W (starting at the prime meridian). A more accurate position is 29°58'N, 90°07'W, which is read as "twenty-nine degrees fifty-eight minutes north (latitude) and ninety degrees seven minutes west (longitude)."

Ellipse

An *ellipse* looks like an elongated circle. It is the path of a planet in the solar system. You can draw an ellipse by choosing two stationary points represented by nails, and a length of string longer than the distance between the points. If we were drawing a circle, we would need only one point (the center), and a length of string, but an ellipse has two points called "foci" (the plural of "focus").

Figure 5

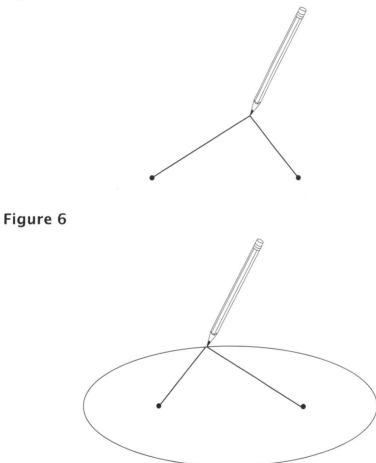

Figure 6

If you put your pencil against the string and move it around the nails, the shape drawn by the pencil will be an ellipse as in figure 6. Because the string has a constant length, we can say that the distances from each focus to a point on the ellipse add up to the same constant length.

Notice also that if a sound or ray of light emanates from one focus, it "bounces off," or is reflected off, the ellipse and arrives at the other focus. This is important in acoustics.

Area of an Ellipse

If an ellipse is drawn on the XY coordinates as in figure 7, then the long, or major, axis would be on the X-axis, and the short, or minor, axis would be on the Y-axis. To find the area of an ellipse, multiply half the length of the short axis times half the length of the long axis times pi (π).

Figure 7

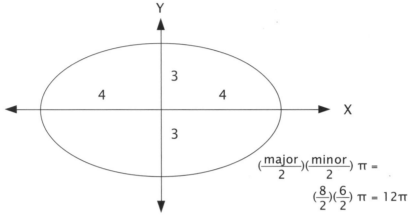

$$\left(\frac{major}{2}\right)\left(\frac{minor}{2}\right) \pi =$$

$$\left(\frac{8}{2}\right)\left(\frac{6}{2}\right) \pi = 12\pi$$

Area = (3)(4)(π) = (12)(3.14) = 37.68 units^2

Finding the area of an ellipse is very similar to finding the area of a circle. However, there is no formula for the distance around an ellipse comparable to that for the circumference of a circle.

Volume of Rectangular Solid and Cylinder

The formula for most volume problems, whether for a cube, a rectangular solid, or a cylinder, is area of the base times the height. To distinguish the volume formula from the area of a rectangle or parallelogram, use a capital B for area of the Base ($V = Bh$). To find the area of a rectangle or a parallelogram, use a lowercase b for the base ($A = bh$).

Volume of a Rectangular Solid

In this lesson, there are several new terms that are used to define the shapes. In figure 1, the flat surface rectangles that make up the rectangular solid are called *faces*. The lines where the faces meet are called *edges,* and the points where the edges meet are called *vertices*. This solid has six faces, twelve edges, and eight vertices. In figure 1, the area of the Base (or basement, if you picture a hotel) is 3 x 4, or 12 (rooms in the basement). The height (or number of floors) is four. The answer is expressed as units cubed, or cubic units.

Figure 1

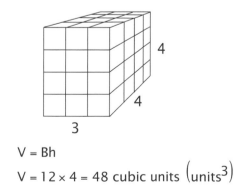

$V = Bh$

$V = 12 \times 4 = 48$ cubic units $\left(\text{units}^3\right)$

Volume of a Cube

A *cube* is a rectangular solid with all edges the same length and all the faces squares. Use the same formula for volume as you did for a rectangular solid.

Example 1

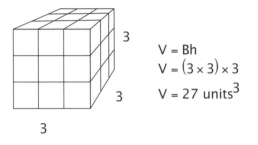

V = Bh
V = (3 × 3) × 3
V = 27 units³

Volume of a Cylinder

What is unique about the area of the base of a *cylinder* is that it is a circle, and therefore the formula for the area of the base is πr². When you find the area of the base, multiply it by the height to find the volume: V = Bh.

I like to think of finding the volume of a cylinder as finding how many pineapple rings are in a can. First you find the area of one pineapple ring (πr²), and then multiply this by the number of pineapple rings (height).

Figure 2

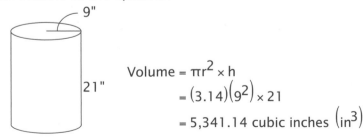

Example 2

Find the volume of the cylinder.

$$\text{Volume} = \pi r^2 \times h$$
$$= (3.14)(9^2) \times 21$$
$$= 5{,}341.14 \text{ cubic inches } (\text{in}^3)$$

Example 3

Find the volume of the cylinder.

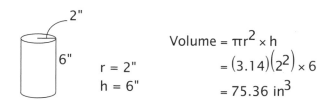

r = 2"
h = 6"

Volume = $\pi r^2 \times h$

$= (3.14)(2^2) \times 6$

$= 75.36$ in^3

Example 4

Find the volume of the cylinder.

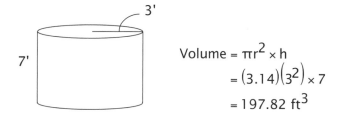

Volume = $\pi r^2 \times h$

$= (3.14)(3^2) \times 7$

$= 197.82$ ft^3

Example 5

Find the volume of the cylinder.

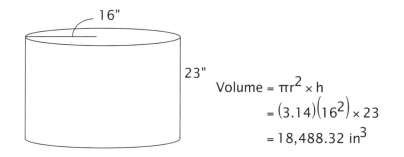

Volume = $\pi r^2 \times h$

$= (3.14)(16^2) \times 23$

$= 18,488.32$ in^3

Volume: Pyramid, Cone, Prism, and Sphere

Volume of a Pyramid and Cone

A pyramid has three or four triangular faces, depending on how many sides in the base. In most of our work, we'll be using pyramids with square bases and four triangular faces. The height of the pyramid itself is the *altitude*. The height of a single face is the *slant height*. The point where the faces meet on top is the *vertex*. The volume of a *pyramid* or a *cone* is found by multiplying B (area of base) times the height (altitude) times one-third: $V = 1/3\ B\ h$.

Figure 1

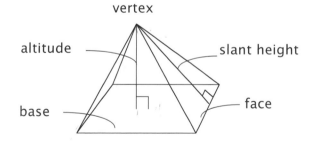

Figure 2

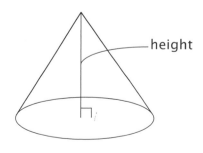

Example 1

Find the volume of the pyramid with a square base.

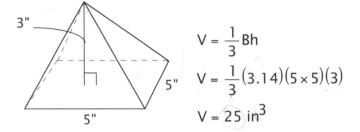

$$V = \frac{1}{3}Bh$$

$$V = \frac{1}{3}(3.14)(5 \times 5)(3)$$

$$V = 25 \text{ in}^3$$

Example 2

Find the volume of the cone.

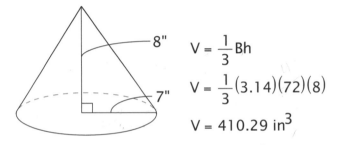

$$V = \frac{1}{3}Bh$$

$$V = \frac{1}{3}(3.14)(72)(8)$$

$$V = 410.29 \text{ in}^3$$

Volume of a Prism

A *prism* has lateral surfaces which are parallelograms and two parallel bases that are also congruent. If the lateral surfaces are perpendicular to the bases, the prism is called a *right prism*. The formula for the volume of a prism is like the formula for the volume of a cylinder: B (area of base) times the height (of the lateral side). In this case the base is a triangle, but it could be a rectangle or other polygon.

Figure 3

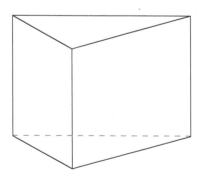

Example 3

Find the volume of the prism.

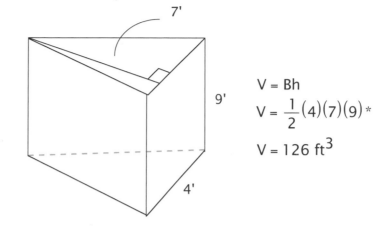

$$V = Bh$$
$$V = \frac{1}{2}(4)(7)(9)*$$
$$V = 126 \text{ ft}^3$$

*Area of the Base is the area of a triangle or 1/2 bh.

Volume of a Sphere

A three-dimensional circle is a **sphere**. The volume of a sphere is 4/3 πr³. The radius goes from the center of the ball to the surface.

Example 4

Find the volume of the sphere.

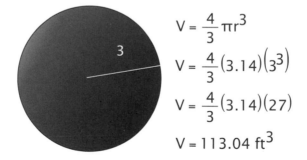

$$V = \frac{4}{3}\pi r^3$$
$$V = \frac{4}{3}(3.14)\left(3^3\right)$$
$$V = \frac{4}{3}(3.14)(27)$$
$$V = 113.04 \text{ ft}^3$$

Surface Area of Solids

Surface Area of Rectangular Solid and Cube

Surface area is the combined area of the outside surfaces of a three-dimensional shape. Take the example of a cardboard box: the surface area of the box is the area of all the sides of the box, including the top and bottom, added together. Surface area differs from volume, which describes in cubic units the amount of space contained inside the box. Surface area is measured in square units. Take apart the box and see how much cardboard makes up the surface area of the box.

Another good way to explain surface area is to observe the room you are in. Count the walls (four) then add the ceiling and floor (two) to get the total number (six) of flat surfaces in the room. In a rectangular solid, these flat surfaces are called *faces*. A *cube*, which is a rectangular solid with all the sides the same length, or a three-dimensional square, also has six faces.

Example 1

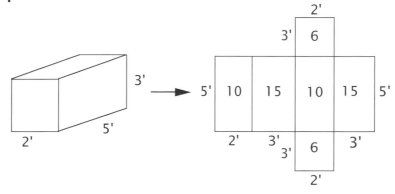

Surface Area = 10 + 10 + 15 + 15 + 6 + 6 = 62 ft^2

Surface Area of a Pyramid

If the base of a pyramid is a square or a rectangle, the pyramid has five flat surfaces. If the base is a triangle, the pyramid has four surfaces. Finding surface area is finding the area of each of the surfaces or faces, and then adding them together.

Example 2

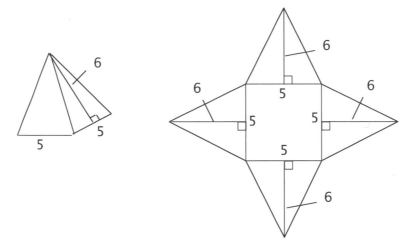

Each of the four triangles is 15 square units (5 x 6 x 1/2), and the base is 25 square units (5 x 5).

Surface Area = [(5 x 6) x 1/2] x 4 + (5 x 5) = 85 units^2

Surface Area of a Cylinder

To find the surface area of a cylinder, find the area of the two bases and add the total to the area of the side. The bases are circles. Picture the area of the side of the cylinder as a rolled up piece of paper that when unrolled becomes a rectangle. One side of the rectangle is the height, and the other is the same as the circumference of the base of the circle.

Example 3

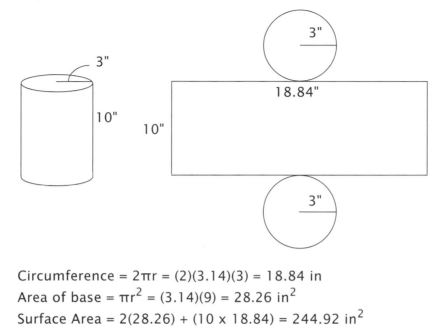

Circumference = $2\pi r$ = (2)(3.14)(3) = 18.84 in
Area of base = πr^2 = (3.14)(9) = 28.26 in^2
Surface Area = 2(28.26) + (10 x 18.84) = 244.92 in^2

Have the student build these shapes out of paper. Below and on the following pages are templates of a cylinder, a rectangular solid, a cube, and two pyramids. Trace the shapes, and then cut the paper along the solid lines. Fold on the dotted lines to form the three-dimensional shapes.

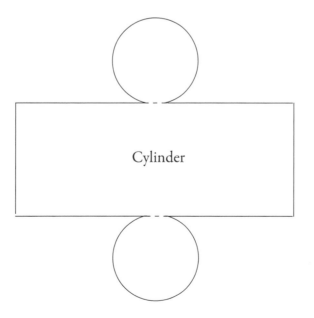

Cylinder

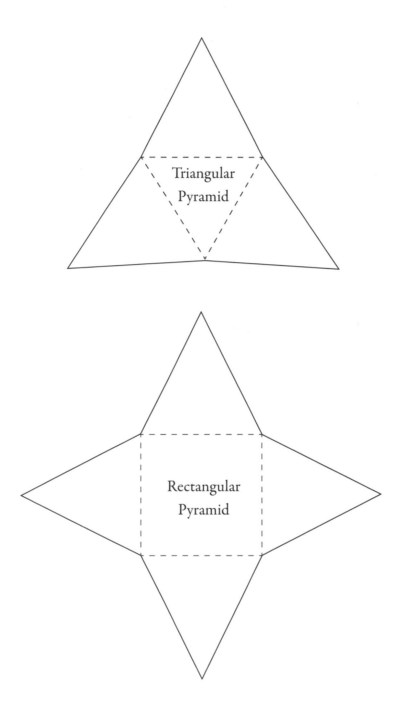

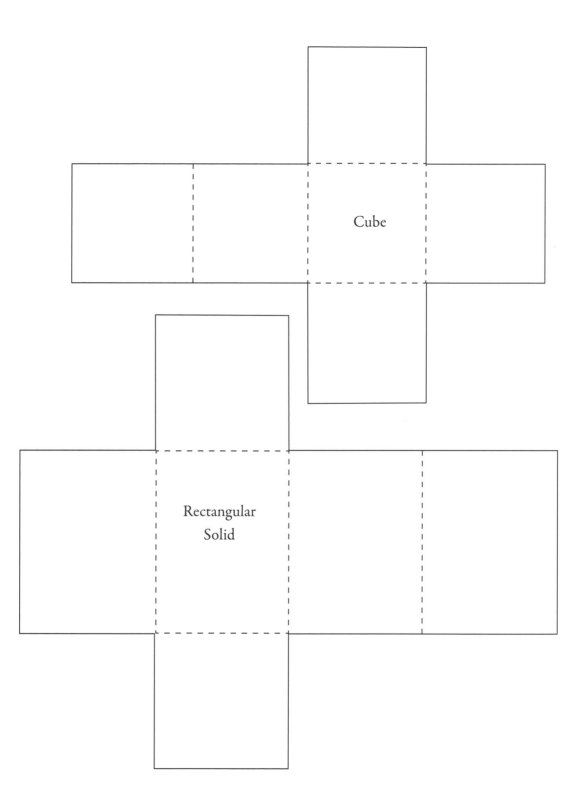

Cube

Rectangular
Solid

Radicals

Adding and Subtracting Radicals

You can only add or combine two things if they are the same kind. $\sqrt{9}$ is the same as three because the square root of nine is three. But $\sqrt{3}$ is not a whole number; it is called a *radical*. You can add a radical to a radical, or a number to a number, but you cannot combine a number and a radical. Just as $2X + 5X = 7X$, $3 + 8 = 11$, and $2X + 5 = 2X + 5$, so $4\sqrt{3} + \sqrt{3} = 5\sqrt{3}$ and $\sqrt{3} + \sqrt{8} = \sqrt{3} + \sqrt{8}$.

However, you can add $2 + \sqrt{9}$ because you can change the square root of nine to a whole number, three, and $2 + 3 = 5$.

Example 1

$$1\sqrt{2} + 2\sqrt{2} = 3\sqrt{2}$$

Just as X is the same as 1X, so $\sqrt{2}$ is the same as $1\sqrt{2}$.

Example 2

$$6\sqrt{5} - 2\sqrt{5} = 4\sqrt{5}$$

Example 3

$$6\sqrt{3} + 2\sqrt{5} = 6\sqrt{3} + 2\sqrt{5}$$

You can't combine these terms, because they aren't the same kind.

Multiplying and Dividing Radicals

You can also multiply and divide radicals. Here are some examples:

Example 4

$$\sqrt{7} \times \sqrt{6} = \sqrt{42}$$

Using what we know of perfect squares, we can estimate that the square root of 7 and the square root of 6 are between the square roots of 4 and 9, so they must be between 2 and 3.

$$\sqrt{4} < \sqrt{6} < \sqrt{7} < \sqrt{9}$$

$$2 < 2.449 < 2.645 < 3$$

The calculator says approximately 2.645 for the square root of 7 and 2.449 for the square root of 6. The answer should be between those two square roots. The calculator approximates 6.48, which is what we expect.

$$\sqrt{6} \times \sqrt{7} \approx (2.449)(2.645) \approx 6.478$$

Or we could estimate the problem of $\sqrt{42}$ using $\sqrt{36}$ and $\sqrt{49}$.

$$\sqrt{36} < \sqrt{42} < \sqrt{49}$$

$$6 < \qquad < 7$$

$\sqrt{42}$ is between 6 and 7 and the approximate root is 6.48.

Example 5

$$2\sqrt{3} \times 4\sqrt{5} = (2 \times 4)(\sqrt{3}\ \sqrt{5}) = 8\sqrt{15}$$

Multiply numbers by numbers, and radicals by radicals.

Example 6

$$\frac{\sqrt{21}}{\sqrt{3}} = \sqrt{7}$$ This is true because $\sqrt{7} \times \sqrt{3} = \sqrt{21}$

This material is not difficult, but it is different. It may take some time to get comfortable with it. The next section on simplifying radicals is perhaps the most difficult of all.

Simplifying Radicals

Simplifying radicals is similar to reducing fractions. You are looking for ways to reduce the large number under the radical sign by bringing out a whole number factor. We can separate a radical into factors. The key is to choose a factor that is a perfect square, such as 4, 9, 16, 25, etc. These are the only factors that may be transformed into whole numbers instead of being left as radicals. In example 7, there are other possible factors, but only $\sqrt{4}$ will become a whole number.

$$\sqrt{4} = 2 \qquad \sqrt{16} = 4 \qquad \sqrt{36} = 6$$

$$\sqrt{9} = 3 \qquad \sqrt{25} = 5 \qquad \sqrt{49} = 7$$

Example 7

$\sqrt{12} = \sqrt{2} \ \sqrt{6} = \sqrt{12}$ This hasn't been simplified.

$\sqrt{12} = \sqrt{4} \ \sqrt{3} = 2\sqrt{3}$ This has been simplified because $\sqrt{4} = 2$

Only perfect squares shed their image and become normal numbers.

Example 8

$\sqrt{18} = \sqrt{3} \ \sqrt{6} = \sqrt{18}$ This hasn't been simplified.

$\sqrt{18} = \sqrt{9} \ \sqrt{2} = 3\sqrt{2}$ This has been simplified because $\sqrt{9} = 3$.

Finding Square Roots with a Calculator

Even though we are treating $\sqrt{2}$ and other radicals as units by themselves, they do represent a value. Even though we can't get an exact value, there are decimal numbers that when squared give you approximately two, three, etc.

Using a calculator, enter two, and then push $\sqrt{}$ the button. The calculator should give you a long decimal number, approximately ($\approx$) 1.41. Next, square this number, and your answer should be approximately two. Check your answers by converting the radicals to decimals, and then approximating the answer.

Example 9

$$\sqrt{18} = \sqrt{9}\ \sqrt{2} = 3\sqrt{2}$$ This has been simplified.

$$4.24 \approx 3.00 \times 1.41$$ This is the same problem using decimal values.

Pythagorean Theorem

Right Triangle

Figure 1 shows a right triangle. There are three sides. The two sides that join to form the right angle are called the *legs*. You can remember this by the letter "L" for "Leg." The letter "L" makes a right angle. Besides the two legs, there is the longest side, which is called the *hypotenuse*; the longest word and the longest side.

Figure 1

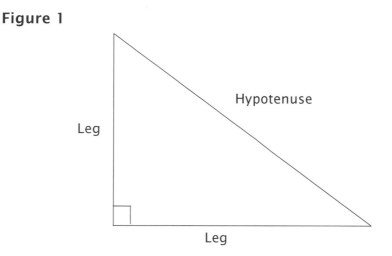

Pythagorean Theorem

The most familiar right triangle is the 3–4–5 right triangle. Figure 2 is a picture of this triangle showing the *Pythagorean theorem*, which is "leg squared plus leg squared equals hypotenuse squared."

Figure 2

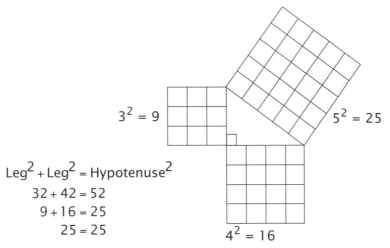

$3^2 = 9$ $5^2 = 25$

$\text{Leg}^2 + \text{Leg}^2 = \text{Hypotenuse}^2$
$3^2 + 4^2 = 5^2$
$9 + 16 = 25$
$25 = 25$

$4^2 = 16$

Converse of Pythagorean Theorem

If you have a right triangle, then the Pythagorean theorem works. The converse is also true: if "leg squared plus leg squared equals hypotenuse squared," then the triangle is a right triangle. Is a 6–8–10 triangle a right triangle? Yes, because $6^2 + 8^2 = 10^2$. $(36 + 64 = 100)$

The examples show how to use this theorem to find the unknown side of a right triangle.

Example 1

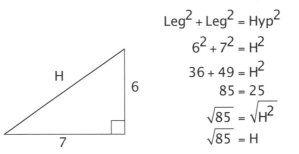

$\text{Leg}^2 + \text{Leg}^2 = \text{Hyp}^2$
$6^2 + 7^2 = H^2$
$36 + 49 = H^2$
$85 = 25$
$\sqrt{85} = \sqrt{H^2}$
$\sqrt{85} = H$

Example 2

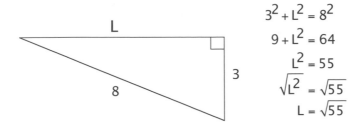

$3^2 + L^2 = 8^2$
$9 + L^2 = 64$
$L^2 = 55$
$\sqrt{L^2} = \sqrt{55}$
$L = \sqrt{55}$

More on Radicals

Radicals in the Denominator

Up to this point, we've been dealing with normal radicals. But there are "radical" radicals that live in places they shouldn't, namely, in the denominator. Only whole numbers are permitted in the denominator. In the example of $\sqrt{2}$, we need to multiply $\sqrt{2}$ by something to make $\sqrt{2}$ a whole number. The easiest factor to choose is $\sqrt{2}$. But we can't randomly multiply the denominator by something, because doing that would change the value of the fraction. If we multiply the numerator by the same factor, then we are multiplying by $\sqrt{2}/\sqrt{2}$, which equals one. Now the radical is in the numerator, which is acceptable, and the denominator is occupied by a whole number, which is also acceptable. Look this over carefully and consider the next four examples.

Example 1

$$\frac{7}{\sqrt{2}} \times \frac{\sqrt{2}}{\sqrt{2}} = \frac{7\sqrt{2}}{\sqrt{4}} = \frac{7\sqrt{2}}{2}$$

Example 2

$$\frac{3}{\sqrt{5}} = \frac{3}{\sqrt{5}} \times \frac{\sqrt{5}}{\sqrt{5}} = \frac{3\sqrt{5}}{\sqrt{25}} = \frac{3\sqrt{5}}{5}$$

Example 3a

$$\frac{4}{\sqrt{8}} = \frac{4}{\sqrt{8}} \times \frac{\sqrt{2}}{\sqrt{2}} = \frac{4\sqrt{2}}{\sqrt{16}} = \frac{4\sqrt{2}}{4} = \frac{\sqrt{2}}{1} = \sqrt{2}$$

Example 3b

You can do example 3 two ways.

$$\frac{4}{\sqrt{8}} = \frac{4}{\sqrt{8}} \times \frac{\sqrt{8}}{\sqrt{8}} = \frac{4\sqrt{8}}{\sqrt{64}} =$$

$$\frac{4\sqrt{4}\sqrt{2}}{8} = \frac{4 \times 2\sqrt{2}}{8} = \frac{8\sqrt{2}}{8} = \sqrt{2}$$

In example 3, it was easier to multiply by $\sqrt{2}$; either way will produce the same answer.

Example 4

$$\frac{3}{\sqrt{2}} + \frac{5}{\sqrt{3}} = \frac{3\sqrt{2}}{\sqrt{2}\sqrt{2}} + \frac{5\sqrt{3}}{\sqrt{3}\sqrt{3}} = \frac{3\sqrt{2}}{2} + \frac{5\sqrt{3}}{3} =$$

$$\frac{3\sqrt{2}}{2} \times \frac{3}{3} + \frac{5\sqrt{3}}{3} \times \frac{2}{2} = \frac{9\sqrt{2}}{6} + \frac{10\sqrt{3}}{6} = \frac{9\sqrt{2} + 10\sqrt{3}}{6}$$

In example 4, first eliminate the radical in the denominator, and then solve by finding the common denominator.

LESSON 20

Special Triangles: 45°-45°-90°

45°-45°-90° Triangle
Have you noticed the relationship between the legs and the hypotenuse of a 45°-45°-90° triangle (right triangle)? Since the angles are congruent, then the opposite sides are also congruent. Think of the triangle shown in figure 1 as half of a square. By putting the same number of slashes on the sides, we indicate that they are congruent.

Figure 1

Find the length of the hypotenuse of the following triangles, and observe the common thread.

Example 1

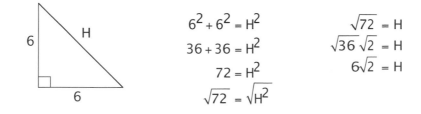

$$6^2 + 6^2 = H^2$$
$$36 + 36 = H^2$$
$$72 = H^2$$
$$\sqrt{72} = \sqrt{H^2}$$

$$\sqrt{72} = H$$
$$\sqrt{36}\,\sqrt{2} = H$$
$$6\sqrt{2} = H$$

Example 2

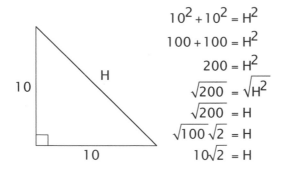

$$10^2 + 10^2 = H^2$$
$$100 + 100 = H^2$$
$$200 = H^2$$
$$\sqrt{200} = \sqrt{H^2}$$
$$\sqrt{200} = H$$
$$\sqrt{100}\sqrt{2} = H$$
$$10\sqrt{2} = H$$

Now let's try a tricky one with A as the length of the legs.

Example 3

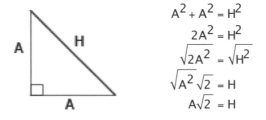

$$A^2 + A^2 = H^2$$
$$2A^2 = H^2$$
$$\sqrt{2A^2} = \sqrt{H^2}$$
$$\sqrt{A^2}\sqrt{2} = H$$
$$A\sqrt{2} = H$$

Rule for Hypotenuse

Have you discovered that the hypotenuse is $\sqrt{2}$ times the leg? If the leg is 5, then the hypotenuse will be $5\sqrt{2}$. If the leg is X, then the hypotenuse is $X\sqrt{2}$.

This rule can also be used to find the length of the legs when the hypotenuse is known.

Example 4

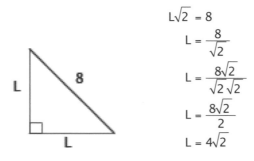

$$L\sqrt{2} = 8$$
$$L = \frac{8}{\sqrt{2}}$$
$$L = \frac{8\sqrt{2}}{\sqrt{2}\sqrt{2}}$$
$$L = \frac{8\sqrt{2}}{2}$$
$$L = 4\sqrt{2}$$

Special Triangles: 30º-60º-90º

30º–60º–90º Triangle

Another special right triangle is the 30º–60º–90º right triangle. As in lesson 20, first we'll try to discover the pattern, and then we'll write our formula.

Example 1

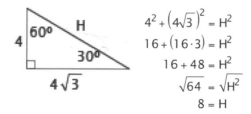

$$4^2 + \left(4\sqrt{3}\right)^2 = H^2$$
$$16 + \left(16 \cdot 3\right) = H^2$$
$$16 + 48 = H^2$$
$$\sqrt{64} = \sqrt{H^2}$$
$$8 = H$$

Example 2

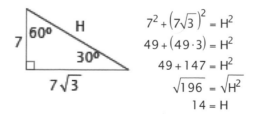

$$7^2 + \left(7\sqrt{3}\right)^2 = H^2$$
$$49 + \left(49 \cdot 3\right) = H^2$$
$$49 + 147 = H^2$$
$$\sqrt{196} = \sqrt{H^2}$$
$$14 = H$$

Now use your observations in examples 1 and 2 to predict the length of the hypotenuse in example 3, and then work it out with the Pythagorean theorem to confirm your hypothesis.

Example 3

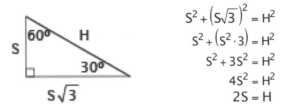

$$S^2 + \left(S\sqrt{3}\right)^2 = H^2$$
$$S^2 + \left(S^2 \cdot 3\right) = H^2$$
$$S^2 + 3S^2 = H^2$$
$$4S^2 = H^2$$
$$2S = H$$

Rule for Hypotenuse

The hypotenuse is twice the length of the short leg (the side opposite the smallest angle). If you are given the short leg, you double it to find the hypotenuse.

Now let's find the relationship between the the short leg (SL) and the long leg (LL). In example 4, we double the short leg to find that the length of the hypotenuse is 10. Our equation to find the long leg is:

Example 4

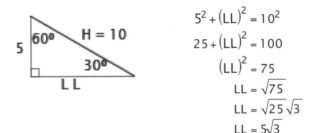

$$5^2 + \left(LL\right)^2 = 10^2$$
$$25 + \left(LL\right)^2 = 100$$
$$\left(LL\right)^2 = 75$$
$$LL = \sqrt{75}$$
$$LL = \sqrt{25}\sqrt{3}$$
$$LL = 5\sqrt{3}$$

Rule for Short Leg

Now let's use variables to confirm our "guesstimate." If the hypotenuse equals 2B, then the short leg equals B.

Example 5

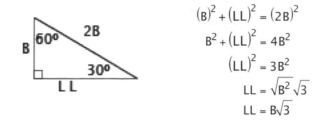

$$(B)^2 + (LL)^2 = (2B)^2$$
$$B^2 + (LL)^2 = 4B^2$$
$$(LL)^2 = 3B^2$$
$$LL = \sqrt{B^2}\sqrt{3}$$
$$LL = B\sqrt{3}$$

Rule for Long Leg

The long leg is $\sqrt{3}$ times the short leg. In example 6, find the length of the short leg and the hypotenuse.

Example 6

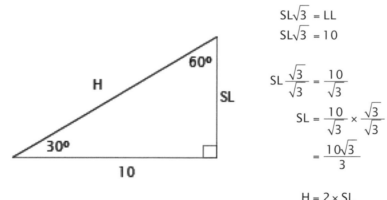

$$SL\sqrt{3} = LL$$
$$SL\sqrt{3} = 10$$

$$SL\frac{\sqrt{3}}{\sqrt{3}} = \frac{10}{\sqrt{3}}$$

$$SL = \frac{10}{\sqrt{3}} \times \frac{\sqrt{3}}{\sqrt{3}}$$

$$= \frac{10\sqrt{3}}{3}$$

$$H = 2 \times SL$$
$$H = 2 \times \frac{10\sqrt{3}}{3}$$
$$H = \frac{20\sqrt{3}}{3}$$

The Converse

If you see that the measures of the angles are 30°–60°–90°, then you know that the length of the sides will follow the pattern of the short leg being one-half of the hypotenuse, and the long leg being the square root of three times the short leg. In short, if the angles are 30°–60°–90°, then the lengths will follow the pattern.

The converse is that if the lengths of the sides follow the pattern of a 30°–60°–90° triangle, the triangle is a 30°–60°–90° triangle. To summarize, if the lengths are "right," then the angles will be "right."

Figure 1

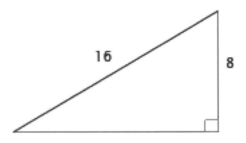

If it is a right triangle and the short leg is one-half the hypotenuse, then you know it is a 30°-60°-90° triangle.

Figure 2

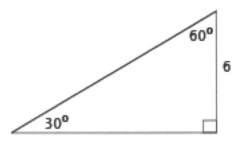

If it is a 30°-60°-90° right triangle, then you know the hypotenuse is two times as long as the short leg (in this case 12).

LESSON 22

Axioms, Postulates, and Theorems

Axioms and *postulates* are observations that one assumes to be true, that make sense, are obvious, but are not readily proven. They are defined as unproved assumptions or assertions. Using axioms or postulates, we make other deductions, and use them to prove new statements true. The result is a **theorem**.

Axioms and postulates are assumed, unproved, and obvious. They are used to validate and prove theorems true. Here is a list of observations that we have found to be true so far.

Postulates and Theorems

1. Vertical angles are congruent.
2. A bisector divides an angle into two congruent angles.
3. A midpoint divides a line segment into two congruent segments.
4. A rectangle has four right angles and two pairs of parallel sides.
5. A parallelogram has two pairs of parallel sides.
6. A square has four right angles, four congruent sides, and two pairs of parallel sides.
7. A rhombus has four congruent sides and two pairs of parallel sides.
8. A trapezoid has one pair of parallel sides.
9. If two parallel lines are cut by a transversal, corresponding angles are congruent.
10. If two parallel lines are cut by a transversal, alternate interior angles are congruent.
11. If two parallel lines are cut by a transversal, alternate exterior angles are congruent.
12. Two angles whose measures add up to 180° are supplementary.

13. Two angles whose measures add up to 90° are complementary.
14. If two angles have equal measures, they are congruent.
15. If two line segments have equal lengths, they are congruent.
16. The measures of the interior angles of a triangle add up to 180°.
17. If a triangle has sides A, B, and C, and A ≤ B ≤ C, then A + B > C.
18. A regular polygon has all sides congruent and all angles congruent.
19. The measures of the exterior angles of a regular polygon add up to 360°.
20. In a right triangle, leg squared plus leg squared equals hypotenuse squared.
21. An isosceles triangle has two congruent sides.
22. Two lines that intersect and form a right angle are perpendicular.
23. Two lines that are coplanar and do not intersect are parallel.
24. New: The property of symmetry: if A = B, then B = A.
25. New: The reflexive property: A = A.
26. New: The transitive property: if A = B and B = C, then A = C.

These postulates and theorems are true as are their converses, or inverses.

Corresponding Parts of Triangles
Remote Interior Angles

Congruent Triangles

If you have two triangles that are congruent—that is, all three angles in the first triangle are congruent to all three angles in the second triangle, and all three sides in the first triangle are congruent to all three sides in the second triangle—then it is important how you write this to reflect which angles and sides *correspond*. Observe the two congruent triangles in figure 1.

Figure 1

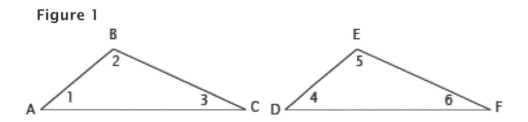

If you were to place triangle ABC on top of triangle DEF, you would notice that point A corresponds to point D, point B corresponds to point E, and point C corresponds to point F. In the same way, $\angle 1$ corresponds to $\angle 4$, $\angle 2$ corresponds to $\angle 5$, and $\angle 3$ corresponds to $\angle 6$. There are several more statements that may be written, and instead of writing them out, I will use symbols. The arrow means "corresponds to."

$$\overline{AB} \leftrightarrow \overline{DE} \qquad \overline{BC} \leftrightarrow \overline{EF} \qquad \overline{CA} \leftrightarrow \overline{FD}$$

$$\triangle ABC \leftrightarrow \triangle DEF \qquad \triangle ACB \leftrightarrow \triangle DFE \qquad \triangle CBA \leftrightarrow \triangle FED$$

Figure 2

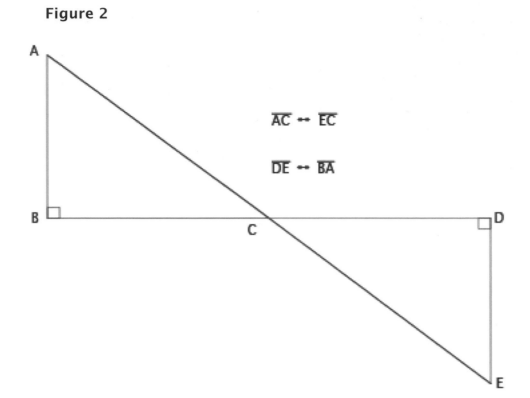

$$\overline{AC} \leftrightarrow \overline{EC}$$

$$\overline{DE} \leftrightarrow \overline{BA}$$

Remote Interior Angles

Note the relationship between ∠1 and ∠2, and ∠3 in example 1 below.

Example 1

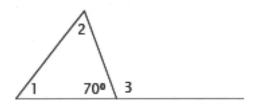

The measure of ∠3 must be 110° because it is supplementary to the 70° angle. Now what would ∠1 plus ∠2 equal? We know that the interior angles of a triangle add up to 180°. So ∠1 + ∠2 + 70° = 180°. Study the following equations that show this:

$$m\angle 1 + m\angle 2 + 70° = 180° \qquad \text{Therefore: } m\angle 1 + m\angle 2 = m\angle 3$$

$$m\angle 3 + 70° = 180° \qquad \text{and } m\angle 1 + m\angle 2 = m\angle 3 = 110°$$

The transitive property states: if $m\angle 1 + m\angle 2 = 110°$, and $m\angle 3 = 110°$, then $m\angle 1 + m\angle 2 = m\angle 3$.

Study the section under example 1 until you are sure you understand this concept. The exterior angle is ∠3. If you are standing at ∠3, then ∠1 and ∠2 are farthest away from you, and these angles are called *remote interior angles*.

Let's look at one more example to make sure this is clear.

Example 2

$$62° + 44° + 74° = 180°$$

$$106° + 74° = 180°$$

$$62° + 44° = 106°$$

We conclude that the measure of the exterior angle of a triangle is equal to the sum of the measures of the two remote interior angles.

Proving Triangles Congruent: SSS and SAS

Congruent Triangles

Two triangles are congruent when all three angles and all three sides of one triangle are congruent to the corresponding parts of the second triangle. In order to prove two triangles are exactly the same, we will be using our powers of observation and common sense, as well as all the postulates and theorems recorded in the previous lessons.

What we will be looking for are different combinations of sides and angles that will function as shortcuts. Instead of finding all six corresponding angles and sides the same before we declare two triangles congruent, perhaps we need find only a few. Consider the angles of a triangle. If we know the measures of two of the angles, we really know the measures of all three. Observe figure 1.

Figure 1

Are all the angles in the first triangle congruent to all the corresponding angles in the second triangle? First find the measures of the two missing angles. We know that the angles in a triangle always add up to $180°$, so the missing angles must each be $40°$. Both of the triangles are $40°–60°–80°$. Earlier, we learned that if we know the measures of two angles, we know the measures of all three, so if two angles of one triangle are congruent to two angles of another triangle, then all three angles are congruent.

Now let's consider the sides of a triangle. Try to build two different triangles with the three sides having lengths of three units, four units, and five units, using the bars. When using the bars, let the inside edges meet at the corners to produce the triangles.

Figure 2

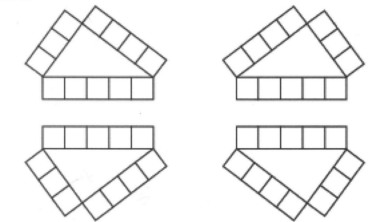

Side-Side-Side Postulate

You may flip the configurations, or twist them in different ways, but you will always end up with the same triangle. Another way of saying that the triangles are the same is to say they are congruent. If you were to measure the angles in all the above triangles, you would find the corresponding angles also are the same. What we've found is that if the three corresponding sides are congruent, it follows that the three corresponding angles are congruent as well. Having learned this, we no longer have to check all six measures (all three sides and all three angles) if we have all three sides. We call this the "*Side-Side-Side*" postulate, or ***SSS***.

There are several combinations of three measures other than SSS that we can use to prove triangles congruent. Try building two triangles, each having a two-bar, a 90° angle, and a three-bar, in that order. Are they congruent?

Figure 3

Side-Angle-Side Postulate

There is no other place to put the hypotenuse, is there? What do you think we call this postulate? Since we use a side, then an included angle, and then another side, the postulate is called **SAS**, for "**Side-Angle-Side.**" Now we'll use these two postulates, SSS and SAS, to prove two triangles congruent.

Proofs

In our proofs, we will always have the following: (1) a figure, or drawing, (2) information given about the figure, and (3) the objective of our proof, or something to prove. As the lawyers in the case, we will do the following: (4) make certain statements, and (5) give reasons to support those statements.

Example 1

(1) **Figure**

(2) **Given:** QRST is a rhombus.

(3) **Prove:** ΔQRS ≅ QTS

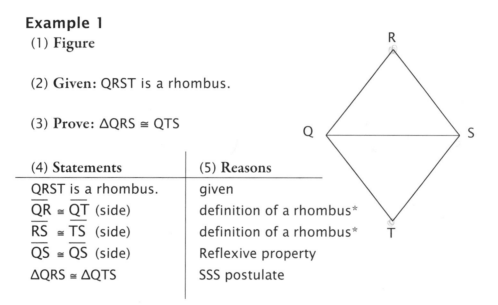

(4) **Statements**	(5) **Reasons**
QRST is a rhombus.	given
$\overline{QR}$ ≅ $\overline{QT}$ (side)	definition of a rhombus*
$\overline{RS}$ ≅ $\overline{TS}$ (side)	definition of a rhombus*
$\overline{QS}$ ≅ $\overline{QS}$ (side)	Reflexive property
ΔQRS ≅ ΔQTS	SSS postulate

*The definition of a rhombus states all the sides are congruent.

Example 2

(1) **Figure**

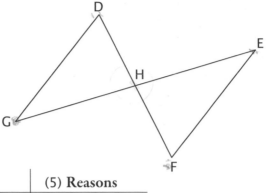

(2) **Given:** H is the midpoint
of $\overline{DF}$ and $\overline{GE}$.

(3) **Prove:** ΔDGH ≅ ΔFEH

(4) **Statements**	(5) **Reasons**
H is the midpoint of $\overline{DF}$ and $\overline{GE}$.	given
$\overline{DH} \cong \overline{FH}$ (S)	H is the midpoint of $\overline{DF}$*
∠DHG ≅ ∠FHE (A)	Vertical angles are congruent
$\overline{GH} \cong \overline{EH}$ (S)	H is the midpoint of $\overline{GE}$*
ΔDGH ≅ ΔFEH	SAS postulate

*The definition of a midpoint states that a midpoint divides a
segment into two congruent segments.

LESSON 25

Proving Triangles Congruent: ASA and AAS
Amplified Parallelogram Theorem

Angle-Side-Angle Postulate

We have more combinations, besides SSS and SAS, for proving triangles congruent. Try a 50° angle, a violet six-bar, and a 70° angle, in that order.

Figure 1

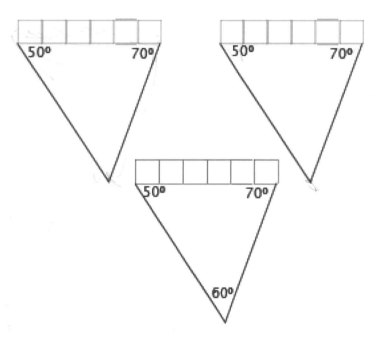

You can see that given an angle, an included side, and another angle, the two remaining sides must intersect at the same point in each triangle. As you've probably figured out, this is the "*Angle-Side-Angle*" or *ASA* postulate.

Angle-Angle-Side Postulate

Now try a 40° angle, a 65° angle, and a brown eight bar, in that order. You really don't have to get out your protractor and your blocks to figure out this one. This is simply another way of stating the ASA postulate, because if you are given two angles, you automatically are given all three. In figure 2, if the first two angles are 40° and 65°, then the third angle must be 75°. In figure 1, which illustrates ASA, the angle where the two lines meet must be 60°, right? So ASA leads to the "*Angle-Angle-Side*" or *AAS* postulate.

Figure 2

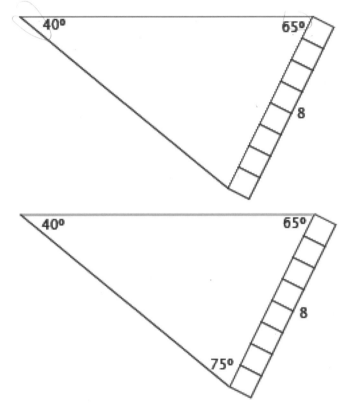

In both of these postulates, when given two angles, you recognize that you also have the third. We could call both of these by one name, AASA. Now that you are becoming adept at proving triangles congruent, remember that you have four postulates to choose from: SSS, SAS, ASA, AAS.

CPCTRC

In lesson 24, we learned that if two triangles, each made up of three angles and three sides, are congruent, then all six corresponding components, or all six corresponding parts, are also congruent. If you are asked to prove two sides or two angles congruent in different triangles, first prove the triangles congruent, then state that the two sides or two angles are also congruent. Give this as your reason: *corresponding parts of congruent triangles are congruent*, abbreviated as *CPCTRC*.

Amplified Parallelogram Theorem (APT)

We have been assuming that opposite sides of a parallelogram are congruent, and you will need to use that fact for some of the proofs in this lesson. Here is a formal proof of the **APT theorem**. If you need to state that two sides of a parallelogram or rectangle are congruent, you may use this theorem. Write "opposite sides of a rectangle are congruent" or "opposite side of a parallelogram are congruent" as your reason. You may also simply write APT.

Example 1

Given: ▱ PLGR is a parallelogram.

Prove: $\overline{PR} \cong \overline{GL}$
$\overline{RG} \cong \overline{LP}$

Statements	Reasons
PLGR is a parallegram.	given
$\overline{PL} \parallel \overline{GR}$, $\overline{PR} \parallel \overline{LG}$	definition of a parallelogram
$\angle LPG \cong \angle RGP$ (A)	Alternate interior angles
$\angle RPG \cong \angle LGP$ (A)	Alternate interior angles
$\overline{PG} \cong \overline{PG}$ (S)	Reflexive property
$\triangle PRG \cong \triangle GLP$	ASA
$\overline{PR} \cong \overline{GL}$, $\overline{RG} \cong \overline{LP}$	CPCTRC

Proving Right Triangles Congruent
HL, LL, HA, and LA

Angles of a Right Triangle

This lesson differs from the two previous lessons in that it applies only to right triangles. The main thing to remember is that of the six components of the two triangles you are seeking to prove congruent, if the triangles are right triangles, one angle in each is already given: the 90° angle. Now you need only one more set of congruent angles to have all three congruent. In a right triangle, if one of the angles is 53°, then you know the other is 37°, because of the 90°, or right angle. With most triangles, you need to know two angles to assure that you have all three congruent. With right triangles, you need only one more angle to assure that you have all three congruent.

Sides of a Right Triangle

Right triangles are also unique because the Pythagorean theorem tells us if we know the measures of two sides, we also know the length of the third side by computing. If we know the lengths of the two legs, we can square them and find the hypotenuse. If we know the lengths of the hypotenuse and one leg, we can square them, then subtract to find the other leg.

The four new combinations for proving right triangles congruent are identical to SSS, SAS, ASA, and AAS discussed in lessons 24 and 25. The postulates have been renamed using right triangle terminology.

Hypotenuse-Leg (HL)

If you are given the hypotenuse and one leg, you can find the length of the other leg, because of the Pythagorean theorem. So **HL** (Hypotenuse-Leg) is the same as SSS.

Leg-Leg (LL)

If you are given the two legs, by using the Pythagorean theorem you can find the length of the hypotenuse. So **LL** (Leg-Leg) is also the same as SSS. But LL could also be SAS because the right angle is between the two legs. We could call it LAL (Leg-Angle-Leg) but it is the same as SAS.

Figure 1

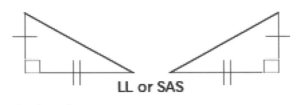

LL or SAS

Hypotenuse-Angle (HA)

If you are given the hypotenuse and one angle (not the right angle), or **HA**, to which of our old postulates is this equivalent? Because you also have the right angle, this is comparable to AAS.

Figure 2

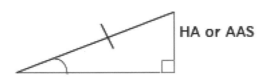

HA or AAS

Leg-Angle (LA)

If you are given the leg and one angle (not the right angle), or **LA**, to which of our old postulates is this equivalent? Because you also have the right angle, this is comparable to either AAS or ASA.

Figure 3

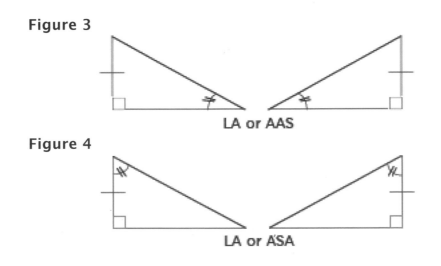

LA or AAS

Figure 4

LA or ASA

Proving Triangles Similar with AA
Proportion or Ratio

Definition of Similar

Think of standing beside a smaller (or larger) picture of yourself. You have the same shape as the picture, but you aren't the same size. Or think of a map of your state. The map isn't the same exact size as the state, but it is in the same proportion. Polygons may have the same proportions, or same shape, without being the same size. Two polygons that are not identical in size but have the same proportions are said to be *similar* (~). Note how the squares in figure 1 are similar but not congruent. The measures of the angles in each square are 90°, but the sides of the squares are different sizes. The squares are similar (~).

Figure 1

In figure 2, we have two triangles that are the same shape and whose corresponding angles have the same measure, but whose corresponding sides are not the same length. (If the angles were the same measure and the sides were the same length, then the triangles would be congruent.)

Figure 2

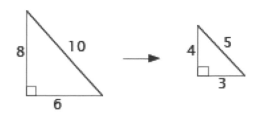

Figure 3

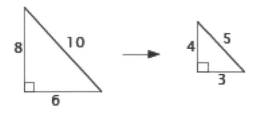

Proportion or Ratio

Figure 3 is the same as figure 2. Looking closely at the corresponding sides, we may write a proportion. You may recall that a proportion is two or more equal ratios. The ratio of the hypotenuse of the small triangle to the hypotenuse of the large triangle is 5 to 10, which may be written as a fraction and subsequently reduced to 1/2.

$$\frac{\text{hyp. small triangle}}{\text{hyp. large triangle}} = \frac{5}{10} = \frac{1}{2}$$

We could also find the ratios of the short legs to the short legs, and the long legs to the long legs, and then put them all together to make a proportion.

$$\frac{\text{short leg small triangle}}{\text{short leg large triangle}} = \frac{3}{6} = \frac{1}{2} \qquad \frac{\text{long leg small triangle}}{\text{long leg large triangle}} = \frac{4}{8} = \frac{1}{2}$$

$$\frac{\text{short leg small triangle}}{\text{short leg large triangle}} = \frac{\text{hyp. small triangle}}{\text{hyp. large triangle}} = \frac{\text{long leg small triangle}}{\text{long leg large triangle}} = \frac{3}{6} = \frac{5}{10} = \frac{4}{8}$$

You also have the option of making a ratio with two parts of one triangle proportional to two parts of the other triangle.

$$\frac{\text{short leg small triangle}}{\text{long leg small triangle}} = \frac{3}{4} \qquad \frac{\text{short leg large triangle}}{\text{long leg large triangle}} = \frac{6}{8} = \frac{3}{4}$$

Angle-Angle Postulate (AA)

If the angles of one triangle are congruent to the corresponding angles of another triangle, but the sides aren't the same, the triangles are similar (~). If the triangles are similar, then the corresponding sides will be proportional. In order for all three sets of angles to be congruent, we really need only two sets of angles to be congruent, and then we automatically have the third set. To prove two triangles similar, we have the *AA* postulate. The *Angle-Angle* postulate states if two angles of one triangle are congruent to two angles of another triangle, then the triangles are similar (~).

Transformational Geometry

Transformational geometry involves moving geometric shapes around and transforming them on a grid. Pretend you've drawn a shape, and then cut it out and moved it from its original position to another location. What we cover in this lesson are four distinct movements that can be used: translation, reflection, rotation, dilation. The first movement is a translation.

Translation

In a *translation*, the shape stays intact and is simply moved to another place on the grid. In example 1, we start with the letter "T" in the second quadrant. The T is moved to the first quadrant. The movement is described in terms of the horizontal (over) and vertical (up or down) coordinates. To measure the movements, pick any point on the shape. I choose a point at the intersection of the two lines in the letter T. Move over six spaces and up two spaces from the chosen point.

Example 1

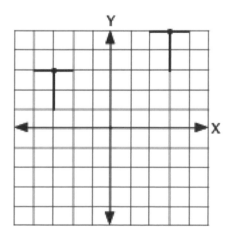

Reflection

Think of a mirror resting on its edge somewhere on the graph. In example 2, we've placed the mirror vertically (running north-south) on the Y-axis. Our figure "R" begins in the second quadrant. I chose two points, A and B, on the R to help in plotting the *reflection* on the graph. The resultant movement is perpendicular to the mirror.

Example 2

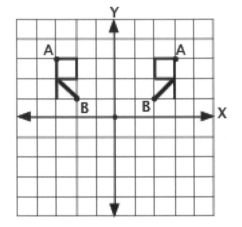

Rotation

When you rotate an object, the rotation occurs around a specific point. Think of your object lying on the edge of a circle with the center of the circle being the point around which you are moving. Since we are dealing with a circle, we'll measure movement by degrees. In example 3, the R has moved 180° around the origin (0, 0). A *rotation* moves counterclockwise around the circle, just as we do when measuring degrees on a graph.

Example 3

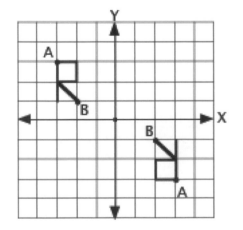

Dilation

When going from darkness into the presence of light, the pupil in the human eye will contract. Conversely, in a dark room, pupils expand (dilate) to allow more light to enter the eye. In the context of transformational geometry, *dilation* is the enlarging or reducing in size of an object without changing its shape. On a computer, you can sometimes click and drag on a corner of an object to change its size without changing its shape. In example 4, our shape is a "D" whose edges are one unit from the origin in each direction. We will enlarge it by a factor of three so the resultant D is three times as large in each direction.

Example 4

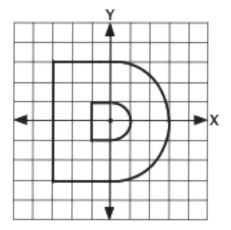

Combining Transformations

You can also combine transformations. In example 5, E moves from the third quadrant to the first quadrant. There are several possibilities of how it got there:

1. Reflection on the Y-axis and a translation up four spaces (figure 1).
2. Reflection on the X-axis and a reflection on the Y-axis (figure 2).
3. Rotation of 180° around the origin (figure 3).

Example 5

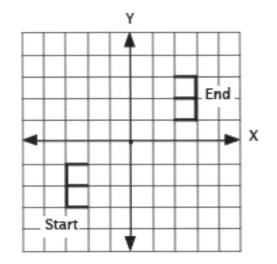

Figure 1

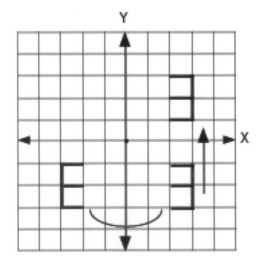

Figure 2

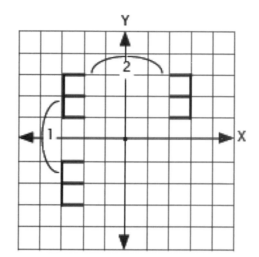

Figure 3

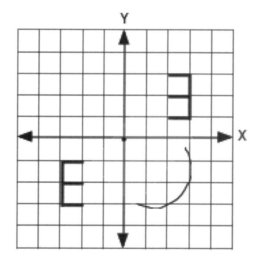

Can you list a different series of transformations that would give the same result as shown in example 5?

Trigonometric Functions
Sine, Cosine and Tangent

Trigonometry

The word **trigonometry** comes from the Greek words τριγονοσ (*trigono*), which means "triangle," and μετρεο (*metreo*), which means "to measure." The Webster's 1828 dictionary defines it as "the measuring of triangles; the science of determining the sides and angles of triangles, by means of certain parts which are given."

Remember that a triangle is composed of three angles and three sides. In trigonometry, unless specified otherwise, we will be dealing with right triangles. In a right triangle, we already know that one of the angles is 90° and that the side opposite the 90° angle is the hypotenuse. Since we are dealing with right triangles, the Pythagorean theorem also applies, and leg squared plus leg squared equals the hypotenuse squared.

Special right triangles such as the 45°–45°–90° and 30°–60°–90° triangles pertain to our study as well. They are taught in lessons 20 and 21. Before proceeding further, make sure the student understands the material in these two lessons, as they lay the foundation for our study of trigonometry.

In order to measure "the sides and angles of triangles by means of certain parts which are given," as our definition tells us, we need to name the angles and sides. For example, in the 30°–60°–90° right triangle, the three sides are referred to as the short leg, the long leg, and the hypotenuse. Notice that the 30° angle (the smallest angle) is opposite the smallest leg, the 60° angle is opposite the longer leg, and the 90° right angle is opposite the hypotenuse. In the 45°–45°–90° triangle, the two legs are congruent so we can't describe them in terms of comparative length. Let's learn a new system of identification to use in trigonometry.

Describing Sides and Angles

We need to describe all the angles and all the sides in figure 1. Since this is a right triangle, we already know that one angle is 90° and that the side opposite the right angle is the hypotenuse. This leaves two angles and two sides, or legs, to name. I've decided to call the angles θ (theta) and α (alpha), both letters in the Greek alphabet. The sides will be described in reference to the angles. If I am standing in angle θ (∠θ), then the leg farthest away from me will be the *opposite* side. The side, or leg, that touches me (where my feet are standing in the illustration) is the *adjacent* side.

Figure 1

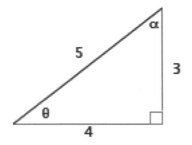

I'll illustrate this with our old friend, the 3–4–5 right triangle. Standing at angle θ, the **opposite** side is three units long and the **adjacent** side is four units long. If I move to angle α, then four is **opposite** and three is **adjacent**. In both instances, the hypotenuse is five.

Trigonometric Ratios

Now we come to the six trigonometric ratios that use the terminology of opposite, adjacent, and hypotenuse. In this lesson we will learn the first three ratios, which are *sine* (sin), *cosine* (cos), and *tangent* (tan). The sine of either angle in a right triangle, in our example angle θ or angle α, is described as the ratio of the opposite side over the hypotenuse.

A fun way to remember these three trigonometric ratios is to think of the result of dropping a brick on your big toe. What would you do? Probably get a pan of water and "soak your toe," or SOH-CAH-TOA.

SOH stands for $\underline{s}$in = $\dfrac{\text{opposite}}{\text{hypotenuse}}$ $\Rightarrow$ $S = \dfrac{O}{H}$

CAH stands for $\underline{c}$os = $\dfrac{\text{adjacent}}{\text{hypotenuse}}$ $\Rightarrow$ $C = \dfrac{A}{H}$

TOA stands for $\underline{t}$an = $\dfrac{\text{opposite}}{\text{adjacent}}$ $\Rightarrow$ $T = \dfrac{O}{A}$

Example 1

Use the 3–4–5 right triangle.

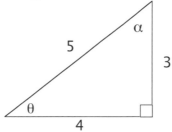

The sine of angle θ is $\dfrac{\text{opposite}}{\text{hypotenuse}} = \dfrac{3}{5}$ or $\sin\theta = \dfrac{3}{5} = .6000$.

The cosine of θ is the $\dfrac{\text{adjacent}}{\text{hypotenuse}}$, or $\cos\theta = \dfrac{4}{5} = .8000$.

And the tangent of θ is the $\dfrac{\text{opposite}}{\text{adjacent}}$, or $\tan\theta = \dfrac{3}{4} = .7500$.

Notice that in a right triangle (which has one right angle), the other two angles always add up to 90°, so they are complementary.

Using the 3–4–5 right triangle, let's look at the trigonometric ratios once again. We have learned that the ratios of the sides are constant for a 45°–45°–90° triangle or a 30°–60°–90° triangle. In the latter, the short side is always one-half the hypotenuse. The length of the sides of each 30°–60°–90° right triangle may vary, but the short side will always be one-half of the hypotenuse. Using our new terminology we can now say:

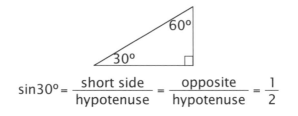

$$\sin 30° = \frac{\text{short side}}{\text{hypotenuse}} = \frac{\text{opposite}}{\text{hypotenuse}} = \frac{1}{2}$$

If the sides of the triangle had lengths such as the following:

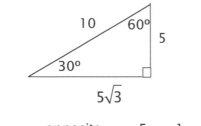

$$\sin 30° = \frac{\text{opposite}}{\text{hypotenuse}} = \frac{5}{10} = \frac{1}{2} = .500$$

or if the side lengths varied and you had:

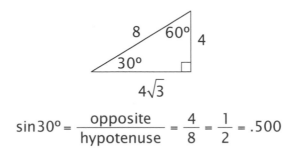

$$\sin 30° = \frac{\text{opposite}}{\text{hypotenuse}} = \frac{4}{8} = \frac{1}{2} = .500$$

Observe that the ratio of the small side to the hypotenuse remains constant.

Example 2

Find the sine, cosine, and tangent ratios for 30° in the triangle.

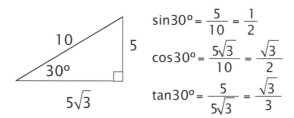

$$\sin 30° = \frac{5}{10} = \frac{1}{2}$$

$$\cos 30° = \frac{5\sqrt{3}}{10} = \frac{\sqrt{3}}{2}$$

$$\tan 30° = \frac{5}{5\sqrt{3}} = \frac{\sqrt{3}}{3}$$

Example 3

Find the sine, cosine, and tangent ratios for 60° in the same triangle.

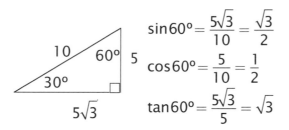

$$\sin 60° = \frac{5\sqrt{3}}{10} = \frac{\sqrt{3}}{2}$$

$$\cos 60° = \frac{5}{10} = \frac{1}{2}$$

$$\tan 60° = \frac{5\sqrt{3}}{5} = \sqrt{3}$$

The ratios $\frac{opp}{hyp}$, $\frac{adj}{hyp}$, and $\frac{opp}{adj}$ depend on what angle is being referred to.

That is why the value of sin 30° and the value of sin 60° are different even though the angles are in the same triangle.

Inverse Trigonometric Functions
Secant, Cosecant, and Cotangent; $Sin^2 + Cos^2 = 1$

Inverse Functions

These are three other ratios that we have in addition to sin = opp/hyp, cos = adj/hyp, and tan = opp/adj.

These ratios are the inverses, or reciprocals, of our original soh-cah-toa.

Where the sine (sin) = $\dfrac{opp}{hyp}$, the **cosecant** (csc) = $\dfrac{hyp}{opp}$.

Where the cosine (cos) = $\dfrac{adj}{hyp}$, the **secant** (sec) = $\dfrac{hyp}{adj}$.

Where tangent (tan) = $\dfrac{opp}{adj}$, the **cotangent** (cot) = $\dfrac{adj}{opp}$.

Now we have our stable full of all the possible trigonometric ratios. Study the following example to clarify what has been explained.

Example 1
Find all six trig ratios for θ.

$$\sin \theta = \frac{3}{5} \qquad \csc \theta = \frac{5}{3}$$

$$\cos \theta = \frac{4}{5} \qquad \sec \theta = \frac{5}{4}$$

$$\tan \theta = \frac{3}{4} \qquad \cot \theta = \frac{4}{3}$$

Sin²θ + Cos²θ = 1

Using the triangle above and what we have learned about trigonometric ratios, $\sin \theta = \frac{O}{H}$ and $\cos \theta = \frac{A}{H}$.

Figure 1

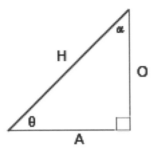

We also know from the Pythagorean theorem that $O^2 + A^2 = H^2$. Let's put these together to develop an important formula that will be used in the further study of trigonometry.

$$O^2 + A^2 = H^2 \qquad \text{Divide through by } H^2.$$

$$\frac{O^2}{H^2} + \frac{A^2}{H^2} = 1 \qquad \text{Simplify the ratios.}$$

$$\left(\frac{O}{H}\right)^2 + \left(\frac{A}{H}\right)^2 = 1 \qquad \text{Replace } \frac{O}{H} \text{ with sin } \theta \text{ and } \frac{A}{H} \text{ with cos } \theta.$$

$$(\sin \theta)^2 + (\cos \theta)^2 = 1 \qquad \text{This is the Pythagorean theorem expressed with trig ratios.}$$

$$\sin^2\theta + \cos^2\theta = 1 \qquad \text{Here is another way of writing the same thing.}$$

Student Solutions

Solutions are shown in detail. The student may use canceling and other shortcuts as long as the answers match. If you see an error, see online solutions mentioned on page 14.

Lesson Practice 1A

1. starting point or origin of a ray
2. contained in the same line
3. having length, but no width
4. an infinite number of connected points
5. ray
6. line segment
7. equal
8. similar
9. congruent
10. false: The common endpoint is B.
11. true
12. true: line BC extends indefinitely in both directions, so it includes $\overline{AB}$.
13. false : They have only one point in common.
14. true: ray BC extends indefinitely to the right, so it includes everything in that direction.
15. = (two quantities)
16. ≅ (two geometric figures)
17. = (two quantities)
18. ≅ (two geometric figures)
19. ≅ (two geometric figures)
20. = (two quantities)

Lesson Practice 1B

1. geometry
2. point
3. line
4. collinear
5. ray
6. segment
7. similar
8. equal
9. congruent

10. endpoint
11. line
12. ray
13. line segment
14. congruent
15. point C
16. ray DE
17. lines AB or BC or AC or BA or CB or CA
18. A or B or C
19. an infinite number
20. rays BA or CA or CB

Systematic Review 1C

1. ray, segment
2. shape, size
3. point
4. line
5. points
6. ray
7. line segment
8. line, congruent to, line
9. ray
10. geometry
11. has same shape but different size
12. exactly the same length or measure
13. in the same line
14. exactly the same shape and size
15. point S
16. rays MP or MQ
17. line RS : Any answer that refers to this line is acceptable.
18. M, P, or Q
19. infinite
20. infinite

Systematic Review 1D

1. line
2. same : both are infinite
3. line segment

4. congruent, equal
5. A
6. measure, earth
7. point
8. similar
9. collinear
10. points
11. line AB is congruent to line CD
12. distance AB is equal to distance CD
13. line segment AB is congruent to line segment CD
14. ray AB is congruent to ray CD
15. false: They do not lie on the same line.
16. true
17. false: They have only one point in common.
18. false: They have no common endpoint.
19. true: They both refer to the same line segment.
20. true: The line is not drawn, but it could be.

Lesson Practice 2A

1. length and width
2. two
3. same
4. two-dimensional; three-dimensional
5. meet
6. combined
7. collection or group
8. null
9. plane
10. subset – ⊏
11. null set – ∅
12. union – ∪
13. intersection – ∩
14. true
15. false
    ray BE ∩ ray BF = point B

16. false : The line segments have no intersection, but their union is simply the two segments.
17. true
18. false : Of the points mentioned, only B is a subset of line EF.

Lesson Practice 2B

1. point
2. line
3. plane
4. coplanar
5. set
6. plane
7. empty or null set
8. three
9. A is a subset of B
10. the union of A and B
11. the intersection of A and B
12. the set containing A and B
13. A is an empty set.
14. false: The union is the two segments.
15. false: Only S is contained in the intersection.
16. true
17. true
18. false: Q is not contained in $\overline{RT}$.

Systematic Review 2C

1. plane
2. coplanar
3. collinear
4. similar
5. intersection
6. union
7. congruent
8. set
9. empty or null set
10. equal
11. union
12. null or empty set

13. subset
14. intersection
15. Any answer representing ray DF or any lines or rays that contain $\overline{DE}$ would be correct.
16. Any answer representing line AB would be correct.
17. $\overline{BC}$ or $\overline{CB}$
18. $\overline{BE}$ or $\overline{EB}$
19. empty or null set
20. Any answer representing line FD would be correct.

Systematic Review 2D
1. empty or null set
2. plane
3. endpoint, origin
4. union
5. intersection
6. subset
7. line
8. point
9. congruent
10. two lines in the same plane
11. two points on the same line
12. two squares with different dimensions
13. two squares with same dimensions
14. two measurements with the same value
15. L
16. $\overline{EH}$
17. $\overrightarrow{EL}$ or $\overrightarrow{EH}$
18. one
19. endpoint, origin
20. K, L, G

Systematic Review 2E
1. set
2. $\varnothing$, or $\{\ \}$
3. ray
4. line segment

5. ⊏
6. union
7. intersection
8. collinear
9. E
10. infinite; Although only 5 points are labeled, any plane contains an infinite number of points.
11. yes. Two lines that intersect are in the same plane.
12. infinite; Although only 3 points are labeled, any line contains an infinite number of points.
13. infinite
14. yes. Any two points can be connected by a straight line.
15. no
16. No. It does not lie in plane x (given).
17. ray AE
18. $(2 \times 5) \times 4^2 - 5^2 =$
$(10) \times 16 - 25 =$
$160 - 25 = 135$
19. $42 \times 3 \div (6^2 \div 12) =$
$42 \times 3 \div (36 \div 12) =$
$42 \times 3 \div 3 =$
$126 \div 3 = 42$
20. $28 \div 2^2 + 6^2 =$
$28 \div 4 + 36 =$
$7 + 36 = 43$

Lesson Practice 3A
1. T
2. T
3. B
4. T
5. ∠RTQ or ∠QTR
6. W
7. ∠XWV or ∠VWX
8. W

9. 116° You may want to accept answers that are a degree or two either way for this and similar problems.
10. 24°
11. 90°
12. 75° – check with protractor
13. 95° – check with protractor
14. 170° – check with protractor

11. 101°
12. 12°
13. 30° – check with protractor
14. 25° – check with protractor
15. 90° – check with protractor
16. B
17. ray CD
18. rays DA or DC
19. C
20. ∅

Lesson Practice 3B

1. J
2. G
3. S
4. H
5. ∠CHB or ∠BHC
6. M
7. ∠XMY or ∠YMX
8. M
9. 138°
10. 49°
11. 24°
12. 15° – check with protractor
13. 160° – check with protractor
14. 110° – check with protractor

Systematic Review 3C

1. angles
2. vertex
3. M
4. measure, angle
5. ∠BMC or ∠CMB
6. true
7. false: the intersection of two planes is a line
8. false: the intersection of two lines is a point
9. true
10. 73°

Systematic Review 3D

1. degrees
2. vertex
3. coplanar
4. measure, angle alpha
5. similar
6. 90°
7. 51°
8. 170°
9. 179° – check with protractor
10. 18° – check with protractor
11. 88° – check with protractor
12. infinite
13. 2 – every plane is two-dimensional
14. line CD
15. line EG or any other answer that refers to the same line
16. C
17. x
18. infinite: every plane contains an infinite number of points
19. $\overline{CD}$
20. line CD

Systematic Review 3E

1. true
2. true
3. false: The union is $\overline{EF}$.

4. true
5. false
6. true: While this line is not shown, such a line could be drawn.
7. true
8. true
9. false: They have no common end point.
10. 16°
11. 90°
12. 122°
13. 13° – check with protractor
14. 125° – check with protractor
15. 170° – check with protractor
16. true: commutative property of addition
17. true: commutative property of multiplication
18. false: Division is not commutative.
19. false: Subtraction is not commutative.
20. true: commutative property of addition

Lesson Practice 4A

1. acute
2. obtuse
3. right
4. acute
5. acute
6. 180°, straight
7. 270°, reflex
8. 90°, right
9. reflex
10. acute
11. reflex
12. obtuse

Lesson Practice 4B

1. straight
2. obtuse
3. reflex

4. right
5. reflex
6. 56°, acute
7. 35°, acute
8. 136°, obtuse
9. obtuse
10. acute
11. right
12. obtuse

Systematic Review 4C

1. ∠AEB or ∠BEA
2. ∠α, ∠β, ∠γ, ∠AEB, ∠BEC, ∠ACE, ∠BED, ∠CED, ∠ABE, ∠BCE, ∠AEC or ∠ECD
3. ∠AEC
4. 90°
5. ∠BAE, ∠EDC, ∠AED or ∠ACD
6. E
7. BED or DEB
8. ∠EBC
9. acute
10. acute
11. obtuse
12. collinear
13. earth, measure
14. congruent
15. ∅: null or empty set
16. true
17. false: A point has zero dimensions.
18. false: A plane has two dimensions.
19. true: Any angle between 90° and 180° is obtuse.
20. false: A line segment has definite length.

Systematic Review 4D

1. acute
2. obtuse
3. reflex

4. right
5. 180°
6. PTR or RTP
7. 90°
8. 90°
9. T
10. lines MS, ST or MT
11. right
12. acute
13. obtuse
14. straight
15. acute
16. see drawing
(labeling of lines can be switched)
17. infinite
18. see drawing
19. ∠CEB, ∠BED, ∠DEA, ∠AEC
20. E

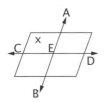

Systematic Review 4E

1. right
2. obtuse
3. acute
4. straight
5. reflex
6. H
7. Z
8. H
9. ∠BHE or ∠EHB
10. ∠YPS or ∠SPY
11. obtuse
12. obtuse
13. straight
14. reflex
15. false: A line is one-dimensional.
16. true

17. false: A point has neither length nor width.
18. true
19.
$$30 = 42Y + 18$$
$$30 - 18 = 42Y$$
$$12 = 42Y$$
$$\frac{12}{42} = Y$$
$$\frac{2}{7} = Y$$

20.
$$15 = -45M - 30$$
$$15 + 30 = -45M$$
$$45 = -45M$$
$$\frac{45}{-45} = M$$
$$-1 = M$$

Lesson Practice 5A

1. parallel
2. perpendicular
3. bisector
4. perpendicular bisector
5. midpoint
6. Follow the procedure in the text. Use a ruler to check that the line segments of each side of the bisector have equal lengths.
7. Follow the procedure in the text. Use a ruler to check that the line segments of each side of the bisector have equal lengths.
8. Follow the procedure in the text. Use a protractor to check that the angles on each side of the bisector have equal measures.
9. Follow the procedure in the text. Use a protractor to check that the angles on each side of the bisector have equal measures.

Lesson Practice 5B

1. right
2. intersect
3. angle, line segment
4. right
5. $\overline{XZ}$
6. Follow the procedure in the text. Use a ruler to check that the line segments of each side of the bisector have equal lengths.
7. Follow the procedure in the text. Use a ruler to check that the line segments of each side of the bisector have equal lengths.
8. Follow the procedure in the text. Use a protractor to check that the angles on each side of the bisector have equal measures.
9. Follow the procedure in the text. Use a protractor to check that the angles on each side of the bisector have equal measures.

Systematic Review 5C

1. c
2. f
3. b
4. e
5. d
6. a
7. E
8. 180°
9. compass, straightedge
10. any angle between 180° and 360°
11. null set
12. Use a ruler to check.
13. Use a ruler to check. The segment on each side of the bisector should measure $1\frac{1}{4}$ in.
14. Use a protractor to check.

15. Use a protractor to check. ∠DEG and ∠FEG should each measure 34°.
16. reflex
17. right
18. acute
19. obtuse
20. straight

Systematic Review 5D

1. bisector
2. parallel
3. reflex
4. perpendicular
5. similar
6. congruent
7. lines QE, QD, ED
8. Q
9. yes
10. Q
11. yes: Although this plane is not shown, any pair of intersecting lines lie in the same plane.
12. Use a ruler to check.
13. Use a ruler to check. The segment on each side of the bisector should measure $1\frac{1}{2}$ in.
14. Use a protractor to check.
15. Use a protractor to check. ∠ABG and ∠CBG should each measure 23°.

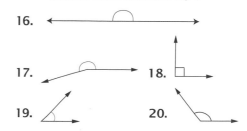

16. 17. 18. 19. 20.

Systematic Review 5E

1. f
2. e
3. b
4. c
5. g
6. a
7. d
8. false: Use a compass and a straightedge
9. true
10. false: The two parts are congruent.
11. false: The line will be perpendicular only if it forms a 90° angle.
12. true
13. Use a ruler to check.
14. Use a ruler to check. The segment on each side of the bisector should measure $\frac{7}{8}$ in.
15. Use a protractor to check.
16. Use a protractor to check. $\angle XYG$ and $\angle ZYG$ should each measure 10°.
17. $24Q + 18Y = 30$
    $6(4Q + 3Y) = 6(5)$
    $4Q + 3Y = 5$
18. $-14Q - 21D = -42$
    $-7(2Q + 3D) = -7(6)$
    $2Q + 3D = 6$
19. $16X - 8 = 56$
    $8(2X - 1) = 8(7)$
    $2X - 1 = 7$
    $2X = 7 + 1$
    $2X = 8$
    $X = \frac{8}{2} = 4$
20. $22X + 33 = 44$
    $11(2X + 3) = 11(4)$
    $2X + 3 = 4$
    $2X = 4 - 3$
    $2X = 1$
    $X = \frac{1}{2}$

Lesson Practice 6A

1. $\angle AHG$, $\angle CHF$
2. $\angle FHB$, $\angle GHD$
3. $\angle AHG$
4. $\angle GHD$
5. $\angle LFK$ or $\angle JFH$
6. $\angle CHA$
7. $\angle HFK$ or $\angle JFL$
8. $\angle DHG$
9. 40°: vertical angles
10. 65°: vertical angles
11. 90°: supplementary angles
12. 50°: complementary angles
13. 115°: supplementary angles
14. 90°: vertical angles
15. f
16. a
17. e
18. b
19. d
20. c

Lesson Practice 6B

1. $\angle MNQ$, $\angle SNR$
2. $\angle MNQ$, $\angle TNP$
3. $\angle YRZ$
4. $\angle TNP$
5. $\angle QNM$ or $\angle PNR$

6. ∠TNP
7. ∠YRZ or ∠SRN
8. ∠SNR
9. 55°: complementary angles
10. 35°: vertical angles
11. 90°: supplementary angles
12. 85°: supplementary angles
13. 40°: vertical angles
14. 55°: vertical angles
15. alpha
16. complementary
17. supplementary
18. gamma
19. vertical
20. delta

Systematic Review 6C

1. 2; 5: If the student referred to these angles using their three-letter names, that would be correct as well.
2. 4
3. BFD
4. BFE or AFD
5. BFD or AFC or AFE
6. 1
7. 40°; complementary angles
8. 40°; If m∠2 = 50°, then m∠1 = 40°, since ∠1 and ∠2 are complementary. If m∠1 = 40°, then m∠4 = 40°, since ∠1 and ∠4 are vertical angles.
9. 1 or 4
10. 140°; supplementary angles
11. any two of angles 1, 2, and 4
12. ∠3; ∠CFE
13. Use a ruler to check. The segments on each side of the bisector should measure $\frac{3}{4}$ in.

14. Use a protractor to check. The angles on each side of the bisector should measure 26°.
15. perpendicular
16. 90°
17. 180°
18. 90°
19. 180°
20. empty or null

Systematic Review 6D

1. true
2. false: They are complementary.
3. true
4. false: Perpendicular angles were not in the list of given information.
5. false: ray GK is the common side.
6. true
7. 39°: vertical angles
8. 51°: complementary angles
9. 90°: perpendicular lines form 90° angles
10. right
11. supplementary
12. 360°
13. f
14. e
15. b
16. a
17. g
18. d
19. h
20. c

Systematic Review 6E

1. lines QR, RV, and QV
2. $\overline{RT}, \overline{XR}, \overline{XT}$
3. 360° √8 = 45°

4. If $m\angle 1 = 90°$, then $m\angle SRV = 90°$ since they are supplementary. $\angle SRV$ is made up of the three smaller angles in the problem, so the sum of their measures is equal to that of $\angle SRV$.

5. obtuse

6. yes: Both are 90°, so they add up to 180°.

7. no: Complementary angles add up to 90°.

8. yes

9. If $\angle$'s 2, 3 and 4 are congruent, and add up to 90°, the measure of each must be $\frac{90°}{3}$ or 30°.
Since $\angle 8$ and $\angle 4$ are vertical angles, they are congruent, so $m\angle 8 = 30°$.

10. 2: vertical angles

11. acute

12. $m\angle 2 + m\angle 3 + m\angle 4 = 90°$
$m\angle 3 = 90° - (25° + 35°)$
$m\angle 3 = 90° - 60° = 30°$

13. $m\angle YRX = m\angle 3$: vertical angles
$m\angle YRX = 30°$ (see #12)

14. ray RQ

15. Use your ruler to check that the resulting line segments are equal in length.

16. Use your protractor to check that the resulting angles are equal in measure.

17. $(-7)^2 = (-7)(-7) = 49$

18. $-(15)^2 = -(15)(15) = -225$

19. $-12^2 = -(12)(12) = -144$

20. $-(9)^2 = -(9)(9) = -81$

Lesson Practice 7A

1. transversal

2. exterior

3. interior

4. congruent

5. alternate

6. parallel

7. same

8. congruent

9. 60°: vertical angles

10. 60°: corresponding angles

11. $\angle 1$ and $\angle 2$ are supplementary, so $m\angle 2 = 180° - m\angle 1 = 180° - 70° = 110°$. $\angle 2$ and $\angle 6$ are corresponding angles, so they are congruent. Thus, $m\angle 6 = 110°$.

12. 70°: corresponding angles

13. 120°: corresponding angles

14. 120°: vertical angles

15. yes: Since $\angle 1$ and $\angle 5$ are corresponding angles, they have the same measure. $\angle$'s 5 and 17 are supplementary, so angles 1 and 17 are also.

16. yes

17. no: They are alternate interior angles.

18. no: They are supplementary angles and add up to 180°. If they were congruent, they would both be 90°.

19. yes: corresponding angles (It may help to ignore line MP.)

20. yes: Angles 12 and 13 are alternate exterior angles. (It may help to ignore lines LR and MP.)

Lesson Practice 7B

1. false

2. true

3. true

4. false: They are always congruent.

5. false: Two parallel lines are cut by a transversal.

6. true

7. true
8. true
9. 110°: alternate interior angles
10. 110°: corresponding angles
11. 85°: corresponding angles
12. 80°: corresponding angles
13. 80°: alternate exterior angles
14. 85°: vertical angles
15. yes: They add up to 180°.
16. yes (It may help to ignore line EF.)
17. no: They are supplementary angles.
18. yes: corresponding angles
19. no: They are corresponding angles, but it is not stated that line AE ‖ line BF.
20. Since they are corresponding angles, if they are congruent, then line AE ‖ line BF.

Systematic Review 7C

1. ∠3; ∠4; ∠5; ∠6
2. ∠1; ∠2; ∠7; ∠8
3. ∠3 and ∠6; ∠5 and ∠4
4. ∠1 and ∠8; ∠2 and ∠7
5. ∠1 and ∠5; ∠3 and ∠7; ∠2 and ∠6; ∠4 and ∠8
6. ∠1
7. 115°
8. They are alternate exterior angles.
9. 115°
10. They are corresponding angles.
11. false: They are supplementary.
12. false: They do not lie on the same line.
13. true
14. true: They are alternate interior angles.
15. The two lines crossed by a transversal are parallel.
16. They are perpendicular.
17. infinite
18. acute
19. obtuse
20. reflex

Systematic Review 7D

1. ‖ or is parallel to
2. 7
3. 3
4. 4
5. 110°: They are supplementary.
6. supplementary
7. vertical
8. if m∠7 = 72°, then
   m∠8 = 180° − 72° = 108°
   if m∠8 = 108°, then
   m∠6 = 108°: corresponding angles
   (Other reasons why may also be correct.)
9. 110°
10. They are alternate exterior angles.
11. true
12. false: It is stated that line RS is not parallel to line VT.
13. true
14. true: vertical angles
15. Corresponding angles are congruent.
16. Their measures add up to 180°.
17. complementary
18. adjacent
19. beta
20. delta

Systematic Review 7E

1. ∠1 and ∠3 are corresponding angles.
   ∠3 and ∠11 are corresponding angles.
   ∠1 ≅ ∠11 so m∠11=100°
2. 100°: They are alternate exterior angles.
3. 80°: ∠1 corresponds to ∠3, ∠3 and ∠4 are supplementary angles.
4. 80°: They are supplementary angles.
5. 13
6. yes

7. yes: ∠7 and ∠2 are
   alternate interior angles.
   ∠2 and ∠10 are
   corresponding angles.
8. no
9. true
10. true: ∠1 and ∠14 are alternate
    exterior angles.
    ∠14 and ∠16 are
    corresponding angles.
11. false
12. true: Parallel lines do not
    intersect.
13. The two lines cut by a
    transversal are parallel.
14. They lie on the same plane.
15. gamma
16. alpha
17. $\dfrac{1}{-1} = -1$
18. $\dfrac{3}{1} = 3$
19. $\dfrac{1}{-2} = -\dfrac{1}{2}$
20. $\dfrac{1}{1} = 1$

Lesson Practice 8A

1. square (or rectangle)
2. rectangle
3. triangle
4. rhombus (or quadrilateral)
5. trapezoid
6. parallelogram (or quadrilateral)
7. $P = 4 + 4 + 4 + 4 = 16$ m
8. $P = 8 + 6 + 8 + 6 = 28$ in
9. $P = 6.1 + 5.5 + 4.9 = 16.5$ ft
10. $P = 10 + 10 + 10 + 10 = 40$ cm
11. $P = 3 + 6.5 + 7 + 8 = 24.5$ in
12. $P = 15 + 23 + 15 + 23 = 76$ mm
13. true
14. true
15. false: They add up to 360º.

16. true
17. false: A right angle is possible
    but not necessary.
18. true
19. false: It has 2 pairs of parallel sides.
20. true

Lesson Practice 8B

1. triangle
2. parallelogram
3. square
4. trapezoid
5. rhombus
6. rectangle
7. $P = 3 + 3 + 3 + 3 = 12$ m
8. $P = 11 + 8 + 11 + 8 = 38$ in
9. $P = 3.9 + 5.0 + 5.3 = 14.2$ ft
10. $P = 18 + 32 + 45 + 23 = 118$ in
11. length of unlabeled horizontal side:
    $4 - 2 = 2$ m
    length of unlabeled vertical side:
    $6 - 4 = 2$ m
    $P = 4 + 4 + 2 + 2 + 2 + 6 = 20$ m
12. length of top horizontal side:
    $40 - 12 - 12 = 16$ in
    $P = 16 + 12 + 12 + 16 + 40 + 16 + 12 + 12$
    $= 136$ in
13. triangle
14. quadrilateral
15. square
16. rhombus
17. triangle
18. quadrilateral
19. trapezoid
20. parallelogram

Systematic Review 8C

1. b
2. a
3. f

4. d
5. e
6. c
7. $P = 3 + 5 + 7 = 15$ m
8. $P = 4 + 6 + 10 + 5 = 25$ in
9. 180°
10. 360°
11. is parallel to, or ||
12. If two corresponding angles are congruent, then the lines are parallel.
13. If two lines are parallel, corresponding angles are congruent.
14. is parallel to, or ||
15. If alternate exterior angles are congruent, the two lines cut by the transversal are parallel.
16. $\angle 12$
17. 106°; $\angle 3$ and $\angle 4$ are supplementary; $\angle 4$ and $\angle 8$ are corresponding.
18. Same as #17, or $\angle 3$ and $\angle 6$ are alternate interior angles; $\angle 6$ and $\angle 8$ are supplementary.
19. 74°
20. $\angle 3$ and $\angle 11$ are corresponding angles.

Systematic Review 8D

1. right
2. quadrilateral
3. square
4. rhombus
5. trapezoid
6. parallelogram
7. $P = 5 + 7 + 11 = 23$ m
8. $P = 4 + 6 + 10 + 5 = 25$ in
9. $P = 10 + 10 + 10 + 10 = 40$ cm
10. length of unlabeled horizontal side:
$10 - 2 - 3 = 5$ in
$P = 2 + 2.5 + 5 + 2.5 +$
$3 + 6 + 10 + 6 = 37$ in

11. transversal; parallel
There may be alternate explanations for #12, 13, 14.
12. 54°; $\angle a$ and $\angle g$ are alternate interior angles.
13. 54°; $\angle b$ and $\angle d$ are alternate interior angles.
14. 72°; $m\angle d + m\angle g = 108°$, therefore $m\angle 2 = 108°$: alternate interior angles
15. acute
16. 108°
17. supplementary or adjacent
18. $\angle FDE$, $\angle FGE$, $\angle 3$, or $\angle 2$
19. a and b (or d and g)
20. If two lines are perpendicular, they form right angles.

Systematic Review 8E

1. true
2. false: They add up to 180°.
3. true
4. false: It has one pair of parallel sides.
5. true
6. false
7. $P = 5 + 4 + 3 = 12$ in
8. length of unlabeled horizontal side:
$12 - 8 = 4$ in
length of unlabeled vertical side:
$8 - 2 = 6$ in
$P = 8 + 6 + 4 + 2 + 12 + 8 = 40$ in
9. $\overline{QT}$ or $\overline{QR}$ or $\overline{ST}$ or $\overline{SQ}$ or $\overline{RT}$
Every line segment in the drawing cuts through a pair of parallel line segments.
10. $\perp$, or is perpendicular to
11. ||, or is parallel to
12. complementary
13. alternate
14. yes

15. no
16. $90° - 43° = 47°$
17. no
18. If the midpoint of line segment DP is point A, A is the middle point of the line segment.
19. slope = -2; y-intercept = 4
20. slope = 1; y-intercept = -2

Lesson Practice 9A

1. $A = bh = (12.4)(10.6) = 131.4 \text{ ft}^2$
2. $A = \text{average base} \times \text{height} =$
$$(\frac{10+15}{2})(5) = (\frac{25}{2})(5)$$
$$= \frac{125}{2} = 62.5 \text{ ft}^2$$
3. $A = \frac{1}{2}bh = \frac{1}{2}(19)(11) = 104.5 \text{ m}^2$
4. $A = (\frac{9.2+11.8}{2})(7.4) = 77.7 \text{ in}^2$
5. $A = (8)(6) = 48 \text{ in}^2$
6. $A = \frac{1}{2}(4)(6) = 12 \text{ ft}^2$
7. $A = (6)(6) = 36 \text{ m}^2$
8. $A = (\frac{5.6+7.8}{2})(3.5) = 23.45 \text{ in}^2$
9. $A = \frac{1}{2}(5)(4.3) = 10.75 \text{ cm}^2$
10. $A = (67)(100) = 6,700 \text{ cm}^2$
11. $A = (2.1)(4.5) = 9.45 \text{ ft}^2$
12. $A = \frac{1}{2}(7)(3) = 10.5 \text{ ft}^2$
13. base, height
14. average
15. half

Lesson Practice 9B

1. $A = (7.4)(4.75) = 35.15 \text{ in}^2$
2. $A = (\frac{9.2+11.8}{2})(7.4) = 77.7 \text{ in}^2$

3. $A = \frac{1}{2}(9.2)(5.5) = 25.3 \text{ m}^2$
4. $A = (\frac{12+16}{2})(8) = 112 \text{ in}^2$
5. $A = (\frac{9+13}{2})(3) = 33 \text{ ft}^2$
6. $A = \frac{1}{2}(3.3)(5.5) = 9.075 \text{ ft}^2$
7. $A = (.05)(.05) = .0025 \text{ m}^2$
8. $A = (\frac{112+156}{2})(70) = 9,380 \text{ ft}^2$
9. $A = \frac{1}{2}(5.33)(3.5) = 9.3275 \text{ ft}^2$
10. $A = (2.33)(1.2) = 2.796 \text{ in}^2$
11. $A = (4)(10) + (2)(3) = 40 + 6 = 46 \text{ in}^2$
12. $A = \frac{1}{2}(28)(12) = 168 \text{ ft}^2$
13. perpendicular
14. trapezoid
15. rectangle

Systematic Review 9C

1. $A = \frac{1}{2}(5)(6) = 15 \text{ cm}^2$
2. $A = (\frac{13+21}{2})(12) = 204 \text{ in}^2$
3. $A = (7)(6) = 42 \text{ ft}^2$
4. $A = (1.5)(4.5) = 6.75 \text{ in}^2$
5. rectangle, square
6. parallelogram, rectangle, square, rhombus
7. square
8. trapezoid
9. a quadrilateral with two pairs of parallel sides
10. a quadrilateral with two pairs of parallel sides and four congruent sides and four right angles
11. yes: corresponding angles
12. yes: alternate interior angles

13. yes: $\angle 5 \cong \angle 7$ because they are corresponding angles, and $\angle 7 \cong \angle 15$ because they are also corresponding angles.

14. 84°: alternate interior angles

15. 96°: $m\angle 10 = m\angle 5$: alternate interior angles. $m\angle 14 = 180° - 84° = 96°$: supplementary angles

    $m\angle 11 = m\angle 14$: alternate interior angles

    (There are several ways to find the answer.)

16. 84°: $m\angle 5 = m\angle 10$: alternate interior angles

    $m\angle 12 = m\angle 10$: corresponding angles

17. 105°: $m\angle 13 = m\angle 5$: corresponding angles

    $m\angle 14 = 180° - 75°$: supplementary angles

18. 75°: corresponding angles

19. 75°: alternate exterior angles

20. length of base:

    $7 + 3 = 10$ in

    length of unlabeled vertical side:

    $6 - 4 = 2$ in

    $P = 3 + 6 + 10 + 4 + 7 + 2 = 32$ in

8. false: It has 2 pairs of parallel sides. If a parallelogram contains one right angle, then it will contain four right angles, and will be a rectangle, which is a special kind of parallelogram.

9. true

10. false: They add up to 360°.

11. yes: They are all 90°, because it is given that the lines are perpendicular.

12. 79°: vertical angles

13. 79°: corresponding angles

14. $m\angle 3 = 180° - 79° = 101°$: supplementary angles

15. yes

16. $\angle 16$

17. $\angle 6$ and $\angle 9$; $\angle 5$ and $\angle 10$; $\angle 7$ and $\angle 12$; or $\angle 8$ and $\angle 11$

18. $\angle 1$ and $\angle 14$; $\angle 2$ and $\angle 13$; $\angle 4$ and $\angle 15$; or $\angle 3$ and $\angle 16$

19. vertical

20. If a quadrilateral is a trapezoid, it has only one pair of parallel sides.

Systematic Review 9D

1. $A = \frac{1}{2}(8)(8) = 32$ cm^2

2. $A = (\frac{6.5 + 10.5}{2})(6) = 51$ in^2

3. $A = (11)(12) = 132$ ft^2

4. $A = (5.1)(2.2) = 11.22$ in^2

5. true

6. true

7. false: Reflex angles measure between 180° and 360°. A 175° angle is obtuse.

Systematic Review 9E

1. $A = (\frac{1}{2})(8)(4) = 16$ cm^2

2. $P = 8 + 7.8 + 5 = 20.8$ cm

3. $A = (\frac{7 + 13}{2})(3) = 30$ in^2

4. $P = 5 + 7 + 13 + 3.5 = 28.5$ in

5. length of unlabeled horizontal side: $14 - 5 - 5 = 4$ cm

    $A = (3)(5) + (3)(4) + (14)(5) = 15 + 12 + 70 = 97$ cm^2

6. $P = 14 + 8 + 4 + 3 + 5 + 3 + 5 + 8 = 50$ cm

7. 90°: It is given that $\overleftrightarrow{MP}$ is perpendicular to $\overleftrightarrow{LN}$.

8. $m\angle 1 + m\angle 3 + m\angle 4 = 180°$
   because they are the three angles
   of a triangle. Since $m\angle 4 = 90°$,
   $m\angle 1 + m\angle 3 + 90° = 180°$,
   or $m\angle 1 + m\angle 3 = 90°$.

9. bisector

10. perpendicular bisector

11. find the average base

12. congruent

13. 360°

14. 360°

15. check with ruler: line
    segment should measure $3\frac{1}{2}$ in

16. check with ruler:
    line segment on each side of the
    bisector should measure $1\frac{3}{4}$ in

17. $A = (2X)(2X) = 4X^2$ units^2
    $P = 2X + 2X + 2X + 2X = 8X$ units

18. $A = (A)(2A) = 2A^2$ units^2
    $P = A + 2A + A + 2A = 6A$ units

19. $A = \frac{1}{2}bh = \frac{1}{2}(A)(B) = \frac{AB}{2}$ or $\frac{1}{2}AB$ units^2
    $P = A + B + C$ units

20. $A = (\frac{4X + 6X}{2})(2X) = (\frac{10X}{2})(2X)$
    $= (5X)(2X) = 10X^2$ units^2
    $P = (4X) + (2X + 2) + (6X) + (2X + 1)$
    $= 14X + 3$ units

Lesson Practice 10A

1. isosceles
2. scalene
3. isosceles
4. right
5. acute
6. obtuse
7. yes
8. the angles would be 90°, 45°, and 45°.
9. no: $5 + 7 < 15$

10. yes: $8 + 9 > 11$
11. isosceles
12. equilateral
13. angles
14. acute
15. obtuse
16. scalene
17. right
18. Triangles will vary.
    One angle must = 90°.
19. Triangles will vary.
    All angles must < 90°.
20. Triangles will vary. Two angles
    must have the same measure.

Lesson Practice 10B

1. equilateral
2. scalene
3. isosceles
4. obtuse
5. right
6. equiangular
7. no
8. In a right triangle, one angle is
   90°, and the other two must each
   be < 90°.
9. yes: $10 + 11 > 12$
10. yes: $2 + 6 > 7$
11. two
12. three
13. zero
14. 90
15. 90
16. must be less than $10 + 8 = 18$
17. obtuse
18. Triangles will vary. All three angles
    must have different measures.
19. Triangles will vary. One angle
    must be > 90°.
20. Triangles will vary. Angles must
    have the same measure of 60°.

Systematic Review 10C

1. isosceles
2. obtuse
3. isosceles, acute
4. triangle will have one 90° and two 45° angles
5. no
6. no : An equilateral triangle is also equiangular. Since the angles are equal and add up to 180°, the measure of each would be $\frac{180°}{3} = 60°$.
7. $A = \frac{1}{2} \times 3\frac{1}{2} \times 2\frac{2}{3} = \frac{1}{2} \times \frac{7}{2} \times \frac{8}{3} = \frac{56}{12} = \frac{14}{3} = 4\frac{2}{3}$ in^2
8. $A = (.05)(.05) = .0025$ m^2
9. $P = 3\frac{1}{2} + 2\frac{2}{3} + 4\frac{2}{5} =$
$\frac{7}{2} + \frac{8}{3} + \frac{22}{5} =$
$\frac{7}{2} \times \frac{15}{15} + \frac{8}{3} \times \frac{10}{10} + \frac{22}{5} \times \frac{6}{6} =$
$\frac{105}{30} + \frac{80}{30} + \frac{132}{30} =$
$\frac{317}{30} =$
$10\frac{17}{30}$ in
10. $P = .05 + .05 + .05 + .05 = .2$ m
11. true
12. false: They add up to 180°.
13. true
14. true
15. false: They are coplanar.
16. true
17. false: The Greek letter beta is β.
18. false
19. true
20. length of unlabeled vertical sides: 5.6 – 3.6 = 2 in
length of unlabeled horizontal side: 1.7 + 1.6 + 2.1 + 1.6 = 7 in
$A = (2)(1.6) + (2)(1.6) + (3.6)(7) = 3.2 + 3.2 + 25.2 = 31.6$ in^2

Systematic Review 10D

1. equilateral
2. acute
3. scalene; obtuse
4. Triangles will vary. One angle must be greater than 90°, and 2 sides must be of equal length.
5. yes
6. A right triangle may have sides of three different lengths.
7. $A = \frac{1}{2}(12)(16) = 96$ cm^2
8. $A = \frac{4+12}{2}(3) = (8)(3) = 24$ m^2
9. $P = 16 + 20 + 12 = 48$ cm
10. $P = 5 + 4 + 5 + 12 = 26$ m
11. yes: alternate interior angles
12. yes: alternate interior angles
13. $m\angle 8 = 90°$
$m\angle 11 = 35°$
$m\angle 7 + m\angle 8 + m\angle 11 = 180°$
$m\angle 7 + 90° + 35° = 180°$
$m\angle 7 = 180° - 90° - 35° = 55°$
14. vertical; right (or supplementary)
15. 180°
16. alpha
17. yes
18. 30°
19. Check with a protractor. The two smaller angles should each measure 20.5°.
20. length of unlabeled vertical sides: 5.6 – 3.6 = 2 in
length of unlabeled horizontal side:
1.7 + 1.6 + 2.1 + 1.6 = 7 in
$P = 7 + 5.6 + 1.6 + 2 + 2.1 + 2 + 1.6 + 2 + 1.7 + 3.6 = 29.2$ in

Systematic Review 10E

1. scalene
2. right
3. equlateral, equiangular

4. All angles should be less than 90°; no angles or sides should have the same measure.

5. no

6. By definition, all angles in an acute triangle are less than 90°.

7. $A = \frac{1}{2}(6)(18.4) = 55.2$ ft^2

8. $A = (7.7)(4.9) = 37.73$ m^2

9. $P = 12 + 10 + 18.4 = 40.4$ ft

10. $P = 7.7 + 5.3 + 7.7 + 5.3 = 26$ m

11. $\perp$; $\angle ACD$ is marked as a right angle

12. bisects; $m\angle 1 = m\angle 2$

13. ACB and ACD, or 1 and 2

14. C is the midpoint of $\overline{BD}$.

15. $A = (11)(11) - (7)(7) =$
$121 - 49 = 72$ in^2

16. $P = 11 + 11 + 11 + 11 = 44$ in

For numbers 17-20,
the last term may vary.

17. $Y = -\frac{1}{2}X - 1$

18. $Y = 3X + 5$

19. $Y = -2X$

20. $Y = 4X + 3$

Lesson Practice 11A

1. c
2. d
3. b
4. e
5. f
6. a
7. 5
8. 6
9. $180° \times 6 = 1{,}080°$
10. $1{,}080° \div 8 = 135°$
11. $180° - 135° = 45°$

12. $45° \times 8 = 360°$

13. $(N - 2)(180°)$

14. dodecagon;
$360°$ total $\div 30° = 12$ sides

15. $8 + 2 = 10$; decagon

16. $(N - 2)(180°) \Rightarrow ((15) - 2)(180°) =$
$13(180°) = 2{,}340°$

17. $2{,}340° \div 15 = 156°$

18. $360° \div 15 = 24°$
for each exterior angle;
$180° - 24° = 156°$
for each interior angle

Lesson Practice 11B

1. b
2. d
3. a
4. f
5. e
6. c
7. 2
8. 3
9. $180° \times 3 = 540°$
10. $540° \div 5 = 108°$
11. $180° - 108° = 72°$
12. $72° \times 5 = 360°$
13. $(N - 2) \times 180°$
14. decagon : $360° \div 36° = 10$ sides
15. Six triangles would mean 8 sides, so it would be an octagon.
16. $(N - 2) \times 180° \Rightarrow ((3) - 2) \times 180° =$
$(1) \times 180° =$
$180°$
17. $180° \div 3 = 60°$
18. Exterior angles add up to $360°$:
$360° \div 3 = 120°$
for each exterior angle.
Interior angles
are $180° - 120° = 60°$.

Systematic Review 11C

1. 3
2. 4
3. $180° \times 4 = 720°$
4. $720° \div 6 = 120°$
5. $180° - 120° = 60°$
6. $60° \times 6 = 360°$
7. square: Exterior angles add up to 360°.
   $360° \div 90° = 4$ sides
8. five sides, so it would be a pentagon
9. $(N - 2)180° \Rightarrow ((12) - 2)180°$
   $= (10)180° = 1,800°$
10. $1,800° \div 12 = 150°$
    check: $360° \div 12 = 30°$
       for each exterior angle.
    $180° - 30° = 150°$
       for each interior angle.
11. 60°: ∠ACB is supplementary to ∠ACD, which has a measure of 90°, so ∠ACB must also have a measure of 90°. ∠ACB, ∠ABC and ∠BAC must add up to 180°, so m∠ABC = $180° - (30° + 90°) = 60°$.
12. ABC : The angles add up to 90°.
13. ∠ADC = 60°, using reasoning similar to that used in question number 11. Since ∠ADC and ∠ADE are supplementary, m∠ADE = $180° -$ m∠ADC = 120°.
14. supplementary
15. equilateral
16. right
17. yes
18. yes: $9 + 8 > 15$
19. A =
    $(1.2)(1.1) + (1.4)(2.2) + (1.1)(3.4) + (1.2)(4.2) =$
    $1.32 + 3.08 + 3.74 + 5.04 = 13.18 \text{ m}^2$
20. P =
    $(1.2 + .8 + 1.1 + 1.2 + 1.4 + 1.1 + 1.2 + 1.1) \times 2 =$
    $9.1 \times 2 = 18.2 \text{ m}$

Systematic Review 11D

1. 4
2. 5
3. $180° \times 5 = 900°$
4. $900° \div 7 \approx 128.57°$
5. $180° - 128.57° = 51.43°$
6. $51.43° \times 7 = 360.01°$
   The .01° is due to rounding in a previous step.
7. hexagon: $360° \div 60° = 6$ sides
8. hexagon
9. $(N - 2)180° \Rightarrow ((9) - 2)180° =$
   $(7)180° = 1,260°$
10. $1,260° \div 9 = 140°$
    check:
    Exterior angles add up to 360°.
    $360° \div 9 = 40°$
       for each exterior angle.
    $180° - 40° = 140°$
       for each interior angle
11. GHK or FHJ
12. JHK
13. yes : They are alternate interior angles. It may help to extend $\overrightarrow{JG}$.
14. yes: They are alternate interior angles. It may help to extend $\overrightarrow{FK}$.
15. isosceles
16. scalene
17. no: $1 + 1 = 2$, and the two short sides need to add up to something greater than the long side.
18. A = bh
19. check with a protractor: angle should measure 125°
20. check with a protractor: new angles should both measure 62.5°

Systematic Review 11E

1. 7
2. 8

3. $180° \times 8 = 1,440°$

4. $1,440 \div 10 = 144°$

5. $180° - 144° = 36°$

6. $36° \times 10 = 360°$

7. triangle: $360° \div 120° = 3$ sides

8. octagon

9. $(N - 2)180° \Rightarrow ((20) - 2)180° =$
$(18)180° = 3,240°$

10. $3,240° \div 20 = 162°$
check: $360° \div 20 = 18°$
$180° - 18° = 162°$

11. $85°$: vertical angles

12. $180° - 85° = 95°$:
supplementary angles

13. $m\angle JFK = 180° - (85° + 45°) =$
$180° - 130° = 50°$

14. $m\angle GJK = 90° - m\angle FJG =$
$90° - 45° = 45°$
The measure of $\angle \alpha$ is
unnecessary for solving
this question.

15. A = average base × height
$A = \dfrac{10 + 17}{2} \times 6 = \dfrac{27}{2} \times \dfrac{6}{1} = \dfrac{162}{2}$
$= 81 \text{ m}^2$

16. $P = 6 + 10 + 11 + 17 = 44 \text{ m}$

17. $Y = X - 1$
$-X + Y = -1$ or
(multiplying both sides by -1)
$X - Y = 1$

18. $2X + Y + 4 = 0$
$Y + 4 = -2X$
$Y = -2X - 4$

19. $Y = 4X + 2$
$-4X + Y = 2$ or
$4X - Y = -2$

20. $X + 2Y - 8 = 0$
$2Y - 8 = -X$
$2Y = -X + 8$
$Y = -\dfrac{1}{2}X + 4$

Lesson Practice 12A

1. sphere

2. circumference

3. chord

4. radius

5. diameter

6. $\overline{GE}, \overline{GC}, \overline{GA},$ or $\overline{GD}$

7. sector

8. arc

9. tangent

10. ellipse

11. perpendicular

12. secant

13. $360° - 60° = 300°$

14. 4

15. $86°$: The measure of an
intercepted arc is the same as
the measure of the central
angle that intercepts it.

16. $86° \div 2 = 43°$: The measure of an
inscribed angle is half the
measure of a central angle
intercepting the same arc.

17. $100°$: Answers that are close
are acceptable.

18. $100°$: Answers that are close are
acceptable, but the answers to
17 and 18 must be the same.

Lesson Practice 12B

1. circumference

2. chord

3. sphere

4. radius

5. radius

6. diameter

7. tangent

8. arc

9. sector

10. two

11. one

12. ellipse
13. $360° - 270° = 90°$
14. 3
15. $44°$
16. $\dfrac{44°}{2} = 22°$
17. $90°$
18. $90°$

19. $900° \div 7 \approx 128.57°$
20. $360° \div 7$ sides $\approx 51.43°$ per exterior angle
    $180° - 51.43° = 128.57°$ per interior angle

Systematic Review 12C

1. $\overline{CB}$ or $\overline{CD}$
2. tangent
3. $\overline{AB}$
4. secant
5. sphere
6. ellipse
7. circumference
8. 5
9. The measure of an inscribed angle is half the measure of the arc that it intercepts, so it would be $10° \times 2 = 20°$.
10. check with a ruler and protractor
11. $44°$: Answers that are close are acceptable.
12. right:
    $\angle 1$ and $\angle 2$ are complementary.
13. yes: $\angle NLP \cong \angle MPL$, and are alternate interior angles.
14. $m\angle 7 = 180° - m\angle 5 = 112°$
    supplementary angles
15. $m\angle 7 = 112°$;
    $m\angle RMN = m\angle 7 = 112°$
    alternate interior angles
16. octagon: $360° \div 45° = 8$ sides
17. quadrilateral: Any answer naming a specific kind of quadrilateral is acceptable.
18. $(N-2)180° \Rightarrow ((7)-2)180° = (5)180° = 900°$

Systematic Review 12D

1. diameter
2. diameter
3. radius
4. secant
5. three
6. ellipses
7. rectangle, square, rhombus, parallelogram
8. circumference
9. inscribed
10. $35° \times 2 = 70°$
11. $360° - 70° = 290°$
12. PLM or MLP
13. vertical angles
14. $m\angle 1 + m\angle 2 = 90°$ (given)
    $m\angle 1 = 90° - 58° = 32°$
    ($m\angle 5$ is unnecessary information)
15. For this problem it may be helpful to ignore everything except $\triangle LMN$. The measures of the angles in this triangle must add up to $180°$:
    $m\angle NLM = m\angle 2 + m\angle 1 =$
    $\quad 58° + 32° = 90°$
    $m\angle 3 = 180° - (m\angle NLM + m\angle 5) =$
    $180° - (90° + 68°) =$
    $180° - 158° = 22°$
16. line segment, line, or ray
17. obtuse angle
18. rhombus
19. scalene triangle
20. octagon

Systematic Review 12E

1. ellipse
2. chord
3. radius
4. diameter or chord
5. A
6. arc
7. sector
8. $\frac{1}{2}$
9. perpendicular
10. Check your drawing using a ruler and a protractor.
11. 225°: Answers that are close to this are acceptable.
12. 5
13. 6
14. $180° \times 6 = 1,080°$
15. $1,080° \div 8 = 135°$
16. $180° - 135° = 45°$
17. $45° \times 8 = 360°$
18. $Y - 2X = 4 \Rightarrow Y = 2X + 4$

    $Y + X = -5 \Rightarrow (2X + 4) + X = -5$

    $\qquad\qquad\qquad 3X + 4 = -5$

    $\qquad\qquad\qquad 3X = -9$

    $\qquad\qquad\qquad X = -3$

    $Y + X = -5 \Rightarrow Y + (-3) = -5$

    $\qquad\qquad\qquad Y = -2$

    $\text{solution} = (-3, -2)$

19. $Y - 4X = 4 \Rightarrow Y = 4X + 4$

    $Y + 2X = -2 \Rightarrow (4X + 4) + 2X = -2$

    $\qquad\qquad\qquad 6X + 4 = -2$

    $\qquad\qquad\qquad 6X = -6$

    $\qquad\qquad\qquad X = -1$

    $Y - 4X = 4 \Rightarrow Y - 4(-1) = 4$

    $\qquad\qquad\qquad Y + 4 = 4$

    $\qquad\qquad\qquad Y = 0$

    $\text{solution} = (-1, 0)$

20. $Y - X = 0 \Rightarrow Y = X$

    $Y - 3X = -6 \Rightarrow (X) - 3X = -6$

    $\qquad\qquad\qquad -2X = -6$

    $\qquad\qquad\qquad X = \frac{-6}{-2} = 3$

    $Y - X = 0 \Rightarrow Y - (3) = 0$

    $\qquad\qquad\qquad Y = 3$

    $\text{solution} = (3, 3)$

Lesson Practice 13A

1. radius
2. circumference
3. $C = \pi d$ or $C = 2\pi r$
4. $A = \pi r^2$
5. x, y, π (or short axis, long axis, π)
6. square
7. latitude
8. longitude
9. minutes
10. prime meridian
11. $C = 2\pi r \approx (2)(3.14)(3) = 18.84$ in
12. $A = \pi r^2 \approx (3.14)(3^2) = 28.26$ in^2
13. $A = \pi r^2 \approx \frac{22}{7}(7^2) = \frac{22}{7}(49) = 154$ m^2
14. $A = \frac{1}{2}(12) \times \frac{1}{2}(8) \times \pi \approx$

    $(6)(4)(3.14) = 75.36$ ft^2
15. 50°7' N; 8°41' E
16. 18°58' N; 72°50' E
17. 4,082 mi
18. $4,082 \times 1.6 = 6,531.2$ km

Lesson Practice 13B

1. diameter
2. circumference
3. area
4. length
5. degrees
6. 0; longitude
7. latitude

8. latitude; longitude
9. seconds
10. square
11. $C = \pi d \approx (3.14)(10) = 31.4$ in
12. $A = \pi r^2 \approx (3.14)(5^2) = 78.5$ in^2
13. $C = \pi d \approx \dfrac{22}{7} \times \dfrac{14}{1} = 44$ m
14. $A = \dfrac{1}{2}(10) \times \dfrac{1}{2}(6) \times \pi \approx$
    $(5)(3)(3.14) = 47.1$ ft^2
15. 40° 43' N; 74° 01' W
16. 33° 55' S; 18° 22' E
17. 7,804 mi
18. $7,804 \times 1.6 = 12,486.4$ km

16. complementary: They are formed from perpendicular lines.
17. obtuse
18. straight
19. $C = 2\pi r \approx (2)(3.14)(8) = 50.24$ units
20. triangle:
    $A = \dfrac{1}{2}bh = \dfrac{1}{2}(3.3 + 3.3)(5.5) = 18.15$ ft^2
    semicircle:
    $A = \dfrac{1}{2}\pi r^2$ (half the area of the whole circle)
    $A \approx \dfrac{1}{2}(3.14)(3.3^2) \approx 17.1$ ft^2
    total:
    $A = 18.15 + 17.1 = 35.25$ ft^2

Systematic Review 13C

1. $A = \pi r^2 \approx (\dfrac{22}{7})(\dfrac{7}{2})^2 = 38.5$ in^2
2. trapezoid
3. check with a ruler: diameter should be 7.5 in
   $C = 2\pi r \approx (2)(3.14)(3.75) = 23.55$ ir
4. radius
5. latitude
6. longitude
7. $A = \dfrac{1}{2}(14) \times \dfrac{1}{2}(4) \times \pi \approx$
   $(7)(2)(3.14) = 43.96$ in^2
8. 4: A regular parallelogram is a square.
9. Use a ruler and protractor to check.
10. 320°: It may be easier to measure the acute angle, and subtract that number from 360°.
11. 64° 09' N; 21° 57' W
12. 39° 55' N; 116° 23' E
13. 4,905 mi
14. $4,905 \times 1.6 = 7,848$ km
15. EGD; vertical angles

Systematic Review 13D

1. $C = 2\pi r \approx 2 \times \dfrac{22}{7} \times 3.5 =$
   $2 \times \dfrac{22}{7} \times \dfrac{7}{2} = 22$ in
2. square or rectangle
3. diameter
4. tangent
5. latitude
6. longitude
7. $A = \dfrac{1}{2}(6) \times \dfrac{1}{2}(2) \times \pi \approx$
   $(3)(1)(3.14) = 9.42$ in^2
8. secant
9. $(N - 2)180° \Rightarrow ((20) - 2)180° =$
   $(18)180° = 3,240°$
   $3,240° \div 20 = 162°$
10. $180° - 162° = 18°$ or
    $360° \div 20 = 18°$
11. $(2)(50°) = 100°$
12. $360° - 100° = 260°$
13. 100°
14. 34° 36' S; 58° 27' W
15. 37° 49° S; 144° 58' E

16. $\angle FGE \cong \angle BGC$: vertical angles
$m\angle BGC + m\angle AGB = 90°$:
complementary angles
$m\angle AGB = 90° - 43° = \boxed{47°}$

17. $m\angle EGD = 90° - 43° = 47°$:
complementary angles
$m\angle EGC = 90° + 47° = \boxed{137°}$

18. $m\angle BGC = m\angle FGC - m\angle FGB = 180° - 135° = 45°$

19. $m\angle AGB = m\angle FGB - 90° = 135° - 90° = 45°$
$m\angle EGD = m\angle AGB = 45°$:
vertical angles

20. semicircle:

$$C = \frac{1}{2}(2\pi r) \approx$$

$$\frac{1}{2}(2)(3.14)(3.3) = 10.36 \text{ ft}$$

sides of triangle:
$6.2 + 6.2 = 12.4 \text{ ft}$
total :
$10.36 + 12.4 = \boxed{22.76 \text{ ft}}$

Systematic Review 13E

1. diameter

2. $A = \pi r^2$

3. $\dfrac{22}{7}$

4. 3.14

5. latitude

6. longitude

7. $A = \dfrac{1}{2}(9) \times \dfrac{1}{2}(3) \times \pi \approx$
$(4.5)(1.5)(3.14) = 21.2 \text{ in}^2$

8. 60

9. sphere

10. diameter

11. $A = \pi r^2 \approx (3.14)(3.5)^2 = 38.47 \text{ m}^2$

12. $C = \pi d \approx (3.14)(1.25) = 3.925 \text{ or } 3.93 \text{ cm}$

13. $360° \div 40° = 9$

14. infinite

15. exterior

16. alternate exterior

17. MLK, CEH

18. GHE

19.
$$3Y + 2X = 12$$
$$(2)(4Y - X = 5) \Rightarrow \underline{8Y - 2X = 10}$$
$$11Y \qquad = 22$$
$$Y = 2$$

$3Y + 2X = 12 \Rightarrow 3(2) + 2X = 12$
$$6 + 2X = 12$$
$$2X = 6$$
$$X = 3$$

solution $= (3, 2)$

20.
$$Y - X = -3$$
$$Y - 2X = -4 \Rightarrow \underline{-Y + 2X = \ \ 4}$$
$$X = \ \ 1$$

$Y - X = -3 \Rightarrow Y - (1) = -3$
$$Y = -2$$

solution $= (1, -2)$

Lesson Practice 14A

1. base, height

2. faces

3. squares

4. edges

5. circle

6. cubic

7. vertices

8. $V = (4)(4)(4) = 64 \text{ in}^3$

9. $V = (5)(4)(3) = 60 \text{ ft}^3$

10. $V = Bh = \pi r^2 h \approx$
$(3.14)(10^2)(5) = 1{,}570 \text{ ft}^3$

11. $V = (5)(7)(4) = 140 \text{ ft}^3$

12. $V = Bh = \pi r^2 h \approx$
$(3.14)(10^2)(25) = 7{,}850 \text{ ft}^3$

Lesson Practice 14B

1. g
2. a
3. c
4. b
5. d
6. h
7. f
8. e
9. $V = (20)(30)(10) = 6{,}000 \text{ ft}^3$
10. $V = Bh = \pi r^2 h \approx$
    $(3.14)(15^2)(60) = 42{,}390 \text{ ft}^3$
11. $V = (10)(10)(10) = 1{,}000 \text{ ft}^3$
12. $V = Bh = \pi r^2 h \approx$
    $(3.14)(10^2)(15) = 4{,}710 \text{ ft}^3$

Systematic Review 14C

1. $V = (3)(5)(4) = 60 \text{ m}^3$
2. $V = Bh = \pi r^2 h \approx$
   $(3.14)(4^2)(4) = 200.96 \text{ ft}^3$
3. $V = Bh = \pi r^2 h \approx$
   $(3.14)(6^2)(10) = 1{,}130.4 \text{ ft}^3$
4. $A = (8)(8) = 64 \text{ cm}^2$
5. $A = \pi r^2 \approx (3.14)(6^2) = 113.04 \text{ cm}^2$
6. $A = (3)(5)(\pi) \approx (3)(5)(3.14) =$
   47.1 cm^2
7. $(N-2)180° \Rightarrow ((8)-2)180° =$
   $(6)(180°) = 1{,}080°$ total
   $1{,}080° \div 8 = 135°$ per angle
8. $180° - 135° = 45°$ or $360° \div 8 = 45°$;
   $45°(8) = 360°$
9. $360° \div 10° = 36$ sides
10. sphere
11. $C = 2\pi r \approx (2)(3.14)(6) = 37.68 \text{ in}$
12. $0°$
13. point

14. $A = \dfrac{5+7}{2}(13) = 78 \text{ in}^2$
15. $180° - (65° + 15°) = 180° - 80° = 100°$
16. obtuse
17. $m\angle CLM = m\angle EHL = 115°$:
    corresponding angles
    $m\angle MLK = 180° - m\angle EHL =$
    $180° - 115° = 65°$
    supplementary angles
18. $m\angle ACL = m\angle MLK = 65°$:
    corresponding angles
19. $\angle CED = m\angle EHL = 115°$:
    corresponding angles
20. $m\angle EHG = 180° - m\angle EHL =$
    $180° - 115° = 65°$:
    supplementary angles

Systematic Review 14D

1. $V = (7)(10)(8) = 560 \text{ m}^3$
2. $V = Bh = \pi r^2 h \approx$
   $(3.14)(3.5^2)(3) = 115.40 \text{ ft}^3$
3. $V = Bh = \pi r^2 h \approx$
   $(3.14)(4^2)(7) = 351.68 \text{ in}^3$
4. $A = (5.4)(4.1) = 22.14 \text{ cm}^2$
5. $A = \pi r^2 \approx (3.14)(11^2) = 379.94 \text{ cm}^2$
6. $V = (10)(4)(\pi) \approx$
   $(10)(4)(3.14) = 125.6 \text{ cm}^2$
7. $(N-2)180° \Rightarrow ((10)-2)180° =$
   $(8)(180°) = 1{,}440°$
   $1{,}440° \div 10 = 144°$
8. $180° - 144° = 36°$
   or $360° \div 10 = 36°$;
   $36° \times 10 = 360°$
9. $360° \div 45° = 8$
10. diameter
11. $23°$
12. $0°$
13. plane
14. yes, because $3 + 4 > 5$

15. $m\angle 10 = m\angle BCA = 62°$
definition of bisector

16. $m\angle 10 + m\angle 11 = 62° + 62° = 124°$

17. $m\angle 9 = 180° - (m\angle 10 + m\angle 11) =$
$180° - (124°) = 56°$
supplementary angles

18. $m\angle 9 = 56°$ (from problem 17)
$m\angle 1 = 56°$: corresponding angles

19. $62°$: $m\angle 7 = m\angle BCA$ because they
are alternate interior angles.

20. $m\angle 11 = m\angle 12$: vertical angles
$m\angle 7 = m\angle 11 = 58°$:
alternate interior
or:
$m\angle 7 = m\angle 12 = 58°$: corresponding

Systematic Review 14E

1. $V = (20)(30)(10) = 6{,}000 \text{ m}^3$

2. $V = Bh = \pi r^2 h \approx$
$(3.14)(10^2)(15) = 4{,}710 \text{ ft}^3$

3. $V = Bh = \pi r^2 h \approx$
$(3.14)(5^2)(7) = 549.5 \text{ in}^3$

4. $A = \frac{1}{2}bh = \frac{1}{2}(8)(6) = 24 \text{ in}^2$

5. $A = \pi r^2 \approx (3.14)(4^2) = 50.24 \text{ cm}^2$

6. $A = \frac{1}{2}(20) \times \frac{1}{2}(15) \times \pi \approx$
$(10)(7.5)(3.14) = 235.5 \text{ units}^2$

7. $(N-2)180° \Rightarrow ((6)-2)180° =$
$(4)180° = 720°$
$720° \div 6 = 120°$

8. $180° - 120° = 60°$ or $360° \div 6 = 60°$
$360°$ is always the total
of exterior degrees.

9. $360° \div 60° = 6$ sides

10. tangent

11. $C = 2\pi r \approx (2)(3.14)(15) = 94.2 \text{ in}$

12. $60'$

13. line (or line segment or ray)

14. $P = 25 + 25 + 25 + 25 = 100 \text{ ft}$

15. $180° - (23° + 35°) =$
$180° - 58° = 122°$

16. minor arc = $23°$;
major arc = $360° - 23° = 337°$

17. $\sqrt{16} = 4$

18. $\sqrt{100} = 10$

19. $\sqrt{25} = 5$

20. $\sqrt{144} = 12$

Lesson Practice 15A

1. slant height

2. altitude

3. vertex

4. $\frac{1}{3}$

5. circle

6. congruent, parallel

7. parallelograms

8. $\frac{4}{3}\pi r^3$

9. $V = \frac{1}{3}Bh = \frac{1}{3}(5)(5)(6) = 50 \text{ in}^3$

10. $V = \frac{1}{3}Bh \approx$
$\frac{1}{3}(3.14)(3^2)(11) = 103.62 \text{ in}^3$

11. $V = Bh = \frac{1}{2}(8)(9)(15) = 540 \text{ ft}^3$

12. $V = \frac{4}{3}\pi r^3 \approx$
$\frac{4}{3}(3.14)(2^3) = 33.49 \text{ ft}^3$ In this
and in other problems of this
type, you may ignore small
answer variations caused by
differences in rounding technique.

13. $V = \frac{1}{3}Bh = \frac{1}{3}(10)(10)(40)$
$\approx 1{,}333.33 \text{ in}^3$

14. $V = \frac{1}{3}Bh = \frac{1}{3}\pi r^2 h \approx$
$\frac{1}{3}(3.14)(2^2)(6) = 25.12 \text{ in}^3$

Lesson Practice 15B

1. prism
2. volume; sphere
3. base; height
4. $\dfrac{1}{3}$
5. square
6. altitude
7. $\dfrac{1}{3}$
8. face
9. $V = \dfrac{1}{3}Bh = \dfrac{1}{3}(3.6)(3.6)(4)$

   $= 17.28 \text{ ft}^3$
10. $V = \dfrac{1}{3}Bh \approx$

    $\dfrac{1}{3}(3.14)(4.2^2)(9.7) = 179.09 \text{ ft}^3$
11. $V = Bh = \dfrac{1}{2}(4.8)(3.2)(7.8)$

    $= 59.9 \text{ cm}^3$
12. $V = \dfrac{4}{3}\pi r^3 \approx \dfrac{4}{3}(3.14)\left(1^3\right) = 4.19 \text{ ft}^3$
13. pyramid:

    $V = \dfrac{1}{3}Bh = \dfrac{1}{3}(6)(6)(3.6) = 43.2 \text{ ft}^3$

    rectangular solid:

    $V = (6)(6)(2.7) = 97.2 \text{ ft}^3$

    total:

    $V = 43.2 + 97.2 = 140.4 \text{ ft}^3$
14. cone: $V = \dfrac{1}{3}Bh = \dfrac{1}{3}\pi r^2 h \approx$

    $\dfrac{1}{3}(3.14)(6)^2(14) = 527.52 \text{ in}^3$

    cylinder: $V = Bh =$

    $\pi r^2 h \approx (3.14)\left(6^2\right)(11) = 1,243.44 \text{ in}^3$

    total: $V = 527.52 + 1,243.44$

    $= 1,770.96 \text{ in}^3$

Systematic Review 15C

1. $V = \dfrac{1}{3}Bh = x\dfrac{1}{3} \times \dfrac{5}{2} \times \dfrac{5}{2} \times \dfrac{9}{2} =$

   $\dfrac{225}{24} = 9\dfrac{3}{8} \text{ m}^3$ or

   $V = \dfrac{1}{3}Bh = \dfrac{1}{3}(2.5)(2.5)(4.5) = 9.375 \text{ m}^3$

2. $V = \dfrac{1}{3}Bh = \dfrac{1}{3}\pi r^2 \approx$

   $\dfrac{1}{3}(3.14)\left(4.5^2\right)(8.25) = 174.86 \text{ in}^3$
3. prism:

   $V = Bh = \dfrac{1}{2}(4)(5)(6) = 60 \text{ ft}^3$

   rectangular solid:

   $V = (5)(6)(2) = 60 \text{ ft}^3$

   total:

   $V = 60 + 60 = 120 \text{ ft}^3$
4. $V = \dfrac{4}{3}\pi r^3 \approx \dfrac{4}{3}(3.14)\left(4^3\right)$

   $= 267.95 \text{ in}^3$
5. altitude
6. They are parallel. (or congruent)
7. $V = (20)(20)(20) = 8,000 \text{ in}^3$
8. $A = \pi r^2 \approx (3.14)\left(5^2\right) = 78.5 \text{ ft}^2$
9. $A = \dfrac{1}{2}(2) \times \dfrac{1}{2}(10) \times \pi \approx (1)(5)(3.14)$

   $= 15.7 \text{ ft}^2$
10. circumference
11. $(N-2)180° \Rightarrow \left((10)-2\right)180° =$

    $(8)180° = 1,440°$

    $1,440° \div 10 = 144°$
12. If interior angle is 120°,

    then exterior angles are

    180° – 120° or 60°.

    Exterior angles always add up

    to 360°. 360° ÷ 60° = 6 sides
13. use protractor to check
14. use protractor to check: New

    angles should each measure 62.5°.
15. A radius and tangent that touch

    a circle at the same point are

    always perpendicular to each other.

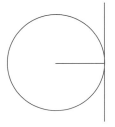

16. $m\angle 6 = m\angle 14 = 50°$:
    corresponding angles
    $m\angle 5 = 180° - m\angle 6 =$
    $180° - 50° = 130°$:
    supplementary angles
    There are other valid ways of
    arriving at this answer.

17. $m\angle BCD = 180° - m\angle 14 =$
    $180° - 50° = 130°$:
    supplementary angles
    $m\angle 10 = \frac{1}{2} m\angle BCD = \frac{1}{2}(130°) = 65°$:
    line AC bisects $\angle BCD$

18. $m\angle 11 = 65°$ from last problem
    $m\angle 4 = m\angle 11 = 65°$:
    corresponding angles
    $m\angle 8 = 180° - m\angle 4 =$
    $180° - 65° = 115°$:
    supplementary angles

19. $m\angle 11 = 65°$ from problem 17
    $m\angle 12 = m\angle 11 = 65°$:
    vertical angles

20. complementary; If two angles are
    complementary, they add up to 90°.

Systematic Review 15D

1. $V = \frac{1}{3} Bh =$

    $\frac{1}{3} \times \frac{13}{2} \times \frac{13}{2} \times \frac{7}{2} = \frac{1,183}{24} = 49\frac{7}{24} \ m^3$
    or
    $\frac{1}{3}(6.5)(6.5)(3.5) \approx 49.29 \ m^3$

2. $V = \frac{1}{3} Bh = \frac{1}{3}\pi r^2 h \approx$

    $\frac{1}{3}(3.14)(3.4^2)(7.6) = 91.96 \ in^3$

3. cone:

    $V = \frac{1}{3} Bh = \frac{1}{3}\pi r^2 h \approx$

    $\frac{1}{3}(3.14)(4^2)(12) = 200.96 \ in^3$

    cylinder:

    $V = Bh = \pi r^2 h \approx (3.14)(4^2)(10) = 502.4 \ in^3$

    total:

    $V = 200.96 + 502.4 = 703.36 \ in^3$

4. $V = \frac{4}{3}\pi r^3 \approx \frac{4}{3}(3.14)(2^3) = 33.49 \ in^3$

5. parallelogram

6. chord

7. $A = bh = (20)(15) = 300 \ in^2$

8. $P = 20 + 20 + 20 + 20 = 80 \ in$

9. $A = \frac{1}{2}(3) \times \frac{1}{2}(6) \times \pi \approx$

    $(1.5)(3)(3.14) = 14.13 \ ft^2$

10. scalene

11. $(N-2)180° \Rightarrow ((8)-2)180° =$
    $(6)180° = 1,080°$
    $1080° \div 8 = 135°$

12. exterior angles $= 180° - 150° = 30°$
    $360° \div 30° = 12$ sides

13. check with ruler

14. check with ruler:

    each half should be $2\frac{3}{8} \ in$

15. check with protractor

16. midpoint

17. $AF = \frac{1}{2} AD = \frac{1}{2} \times 2\frac{7}{8} = 1\frac{7}{16}$

18. $m\angle BFD = m\angle AFE = 88.5°$:
    vertical angles

    $m\angle BFC = \frac{1}{2} m\angle BFD = \frac{1}{2}(88.5°) = 44.25°$:

    bisector

19. $m\angle EFD = 180° - m\angle AFE =$
    $180° - 88.5° = 91.5°$:
    supplementary angles

20. supplementary; If two angles add
    up to 180°, they are supplementary.

Systematic Review 15E

1. $V = Bh = \frac{1}{2}(4.5)(4.2)(11)$

   $= 103.95 \text{ mm}^3$

2. $V = \frac{4}{3}\pi r^3 \approx \frac{4}{3}(3.14)(1.5^3)$

   $= 14.13 \text{ cm}^3$

3. pyramid:

   $V = \frac{1}{3}Bh = \frac{1}{3}(8)(8)(6) = 128 \text{ ft}^3$

   rectangular solid:

   $V = (8)(8)(3) = 192 \text{ ft}^3$

   total:

   $V = 128 + 192 = 320 \text{ ft}^3$

4. altitude

5. prism

6. $\frac{1}{3}$

7. $A = bh = (8)(3.5) = 28 \text{ in}^2$

8. $P = 8 + 4 + 8 + 4 = 24 \text{ in}$

9. $A = \frac{1}{2}(8) \times \frac{1}{2}(10) \times \pi \approx (4)(5)(3.14$

   $= 62.8 \text{ ft}^2$

10. equilateral

11. $(N-2)180° \Rightarrow ((5)-2)180° =$

    $(3)180° = 540°$

    $540° \div 5 = 108°$

12. 8

13. exterior angles =

    $180° - 135° = 45°$;

    $360° \div 45° = 8 \text{ sides}$

14. check with ruler

15. check with protractor:

    angles should measure 55°

16. check with protractor

17. $8^3 \cdot 8^4 = 8^{3+4} = 8^7$

18. $2^8 \div 2^3 = 2^{8-3} = 2^5$

19. $x^2 x^3 y^5 y^{-1} = x^{2+3} y^{5+(-1)} = x^5 y^4$

20. $\frac{1}{10^3} = \frac{10^{-3}}{1} = 10^{-3}$

Lesson Practice 16A

1. 6

2. 5

3. 4

4. circles; rectangle

5. square

6. height; circumference

7. $SA = 2(3)(5) + 2(3)(4) + 2(4)(5) =$

   $30 + 24 + 40 = 94 \text{ ft}^2$

8. $SA = 2\pi r^2 + 2\pi rh \approx$

   $(2)(3.14)(10^2) + (2)(3.14)(10)(5) =$

   $628 + 314 = 942 \text{ ft}^2$

9. $SA = (5)(5) + (4)(\frac{1}{2})(5)(6) =$

   $25 + 60 = 85 \text{ in}^2$

10. base:

    $SA = \frac{1}{2}(6)(5) = 15 \text{ ft}^2$

    sides:

    $SA = (3)\frac{1}{2}(6)(8) = 72 \text{ ft}^2$

    total:

    $SA = 15 + 72 = 87 \text{ ft}^2$

11. $SA = 2(5)(4) + 2(5)(7) + 2(4)(7) =$

    $40 + 70 + 56 = 166 \text{ ft}^2$

12. "roof":

    $SA = 2(4.5)(6) = 54 \text{ ft}^2$

    sides:

    $SA = 2(4)(5) + 2(4)(6)$

    $= 40 + 48 = 88 \text{ ft}^2$

    triangles:

    $SA = 2\frac{1}{2}(4)(5) = 20 \text{ ft}^2$

    bottom:

    $SA = (5)(6) = 30 \text{ ft}^2$

    total:

    $SA = 54 + 88 + 20 + 30 = 192 \text{ ft}^2$

Lesson Practice 16B

1. square

2. pyramid

3. cube

4. triangular

5. cylinder

6. rectangular

7. SA =
$2(10)(30) + 2(10)(20) + 2(20)(30) =$
$600 + 400 + 1,200 = 2,200 \text{ ft}^2$

8. $SA = 2\pi r^2 + 2\pi rh \approx$
$(2)(3.14)(15^2) + (2)(3.14)(15)(60) =$
$1,413 + 5,652 = 7,065 \text{ in}^2$

9. $SA = (3.6)(3.6) + (4)(\frac{1}{2})(3.6)(5.2) =$
$12.96 + 37.44 = 50.4 \text{ ft}^2$

10. SA =
$(2)(\frac{1}{2})(4.8)(3.2) + (2)(4)(7.8) + (4.8)(7.8) =$
$15.36 + 62.4 + 37.44 = 115.2 \text{ cm}^2$

11. SA =
$(2)(\frac{1}{2})(8)(9) + (11)(15) + (8)(15) + (9)(15) =$
$72 + 165 + 120 + 135 = 492 \text{ in}^2$

12. "roof" :
$SA = (2)(6)(6) = 72 \text{ ft}^2$
triangles:
$SA = (2)(\frac{1}{2})(5)(4) = 20 \text{ ft}^2$
sides :
$SA = (2)(5)(5) + (2)(6)(5) = 50 + 60 = 110 \text{ ft}^2$
bottom:
$SA = (5)(6) = 30 \text{ ft}^2$
total:
$SA = 72 + 20 + 110 + 30 = 232 \text{ ft}^2$

Systematic Review 16C

1. $SA = (2)(2) + (4)(\frac{1}{2})(2)(4) = 4 + 16$
$= 20 \text{ m}^2$

2. $SA = 2\pi r^2 + 2\pi rh \approx$
$(2)(3.14)(5^2) + (2)(3.14)(5)(12) =$
$157 + 376.8 = 533.8 \text{ cm}^2$

3. "roof":
$SA = (2)(9)(12) = 216 \text{ m}^2$
triangles:
$SA = (2)(\frac{1}{2})(9)(8) = 72 \text{ m}^2$
sides:
$SA = (2)(6)(12) + (2)(6)(9)$
$= 144 + 108 = 252 \text{ m}^2$
bottom:
$SA = (9)(12) = 108 \text{ m}^2$
total:
$SA = 216 + 72 + 252 + 108 = 648 \text{ m}^2$

4. slant height

5. $V = \frac{4}{3}\pi r^3 \approx \frac{4}{3}(3.14)(1.66^3)$
$\approx 19.15 \text{ in}^3$

6. latitude

7. straight angle

8. $C = 2\pi r \approx (2)(3.14)(7) = 43.96 \text{ ft}$

9. $V = Bh = \pi r^2 h \approx (3.14)(5^2)(12)$
$= 942 \text{ cm}^3$

10. secant

11. $(N - 2)180° \Rightarrow ((12) - 2)180° =$
$(10)180° = 1,800°$
$1,800° \div 12 = 150°$

12. exterior angle $= 180° - 108° = 72°$
$360° \sqrt{72°} = 5 \text{ sides}$

13. check with protractor

14. check with protractor

15. check with ruler and protractor:
all angles should measure 90°

16. rays EF, FE, FB, BF, EB, or BE

17. $\overline{AF}$, $\overline{FD}$, and $\overline{AD}$

18. $m\angle BFC = \frac{1}{2}(m\angle BFD) =$
$\frac{1}{2}(90°) = 45°$:
definition of a bisector
$m\angle AFC = m\angle AFB + m\angle BFC =$
$90° + 45° = 135°$

19. ⊥, or is perpendicular to
20. congruent; If alternate interior
angles are congruent, they are
formed by parallel
lines cut by a transversal.

Systematic Review 16D

1. SA =

$$(2)(\frac{1}{2})(9)(4.2)+(2)(6.2)(11)+(9)(11) =$$

$$37.8+136.4+99 = 273.2 \text{ mm}^2$$

2. $SA = 2\pi r^2 + 2\pi rh \approx$

$$(2)(3.14)(8^2)+(2)(3.14)(8)(10) =$$

$$401.92+502.4 = 904.32 \text{ cm}^2$$

3. triangles:

$$SA = (4)(\frac{1}{2})(6)(5.2) = 62.4 \text{ ft}^2$$

sides:

$$SA = (4)(6)(2.7) = 64.8 \text{ ft}^2$$

bottom:

$$SA = (6)(6) = 36 \text{ ft}^2$$

total:

$$62.4+64.8+36 = 163.2 \text{ ft}^2$$

4. altitude

5. $V = \frac{4}{3}\pi r^3 \approx \frac{4}{3}(3.14)(1.5^3)$

$$= 14.13 \text{ in}^3$$

6. $V = \frac{1}{3}Bh = \frac{1}{3}\pi r^2 h \approx$

$$\frac{1}{3}(3.14)(2.8^2)(4.7) \approx 38.57 \text{ cm}^3$$

7. reflex angle

8. 4

9. $A = \frac{1}{2}(10) \times \frac{1}{2}(20) \times \pi \approx$

$$(5)(10)(3.14) = 157 \text{ in}^2$$

10. sector

11. $(N-2)180° \Rightarrow ((5)-2)180° =$

$$(3)180° = 540°$$

12. $(N-2)(180°) = 360°$

$$180N° -360° = 360°$$

$$180N° = 720°$$

$$N = \frac{720°}{180°} = 4 \text{ sides}$$

13. check with ruler

14. check with protractor:
Fourth angle is 60°.

15. Interior angles of a quadrilateral
add up to 360°, so fourth angle
has a measure of:
$360° - (65°+115°+120°) =$
$360° - 300° = 60°$

16. 90°, because $\overrightarrow{AD} \perp \overrightarrow{EC}$

17. complementary; they add up to 90°

18. supplementary; they form
a straight line

19. $m\angle BGC =$
$m\angle EGC - (m\angle EGF + m\angle FGB) =$
$180° - (34°+90°) =$
$180° - 124° = 56°$

20. perpendicular; If 90° angles
are formed from intersecting
lines, the lines are perpendicular.

Systematic Review 16E

1. SA =

$$(2)(5)(15)+(2)(5)(25)+(2)(15)(25) =$$

$$150+250+750 = 1,150 \text{ in}^2$$

2. $SA = 2\pi r^2 + 2\pi rh \approx$

$$(2)(3.14)(4^2)+(10)(3.14)(8) =$$

$$100.48+251.2 = 351.68 \text{ cm}^2$$

3. "roof":

$$SA = (4)(\tfrac{1}{2})(8)(9) = 144 \text{ ft}^2$$

sides:

$$SA = (4)(8)(3) = 96 \text{ ft}^2$$

bottom:

$$SA = (8)(8) = 64 \text{ ft}^2$$

total:

$$144 + 96 + 64 = 304 \text{ ft}^2$$

4. c

5. d

6. a

7. e

8. h

9. b

10. j

11. f

12. i

13. g

14. measure of minor arc AB =
$m\angle AOB = 40°$

15. measure of minor arc ABC =
$m\angle AOB + m\angle BOC =$
$40° + 100° = 140°$

16. measure of arc ABCD =
$m\angle AOB + m\angle BOC + m\angle COD =$
$40° + 100° + 40° = 180°$

17.
$$\begin{array}{r} X + 4 \\ \times\ X + 8 \\ \hline 8X + 32 \\ X^2 + 4X \\ \hline X^2 + 12X + 32 \end{array}$$

18.
$$\begin{array}{r} X + 5 \\ \times\ X + 2 \\ \hline 2X + 10 \\ X^2 + 5X \\ \hline X^2 + 7X + 10 \end{array}$$

19.
$$\begin{array}{r} X + 3 \\ \times\ X - 1 \\ \hline -X - 3 \\ X^2 + 3X \\ \hline X^2 + 2X - 3 \end{array}$$

20.
$$\begin{array}{r} X - 4 \\ \times\ X + 6 \\ \hline 6X - 24 \\ X^2 - 4X \\ \hline X^2 + 2X - 24 \end{array}$$

Lesson Practice 17A

1. $5\sqrt{2} + 3\sqrt{5} = 5\sqrt{2} + 3\sqrt{5}$:
   cannot be simplified

2. $8\sqrt{7} + 3\sqrt{3}$ cannot be simplified

3. $12\sqrt{6} - 10\sqrt{6} = (12 - 10)\sqrt{6} = 2\sqrt{6}$

4. $11\sqrt{2} + 3\sqrt{2} + 5\sqrt{2} =$
   $(11 + 3 + 5)\sqrt{2} = 19\sqrt{2}$

5. $\dfrac{12\sqrt{24}}{6\sqrt{3}} = \dfrac{12\sqrt{8}}{6} = 2\sqrt{8} =$
   $2\sqrt{4}\sqrt{2} = 2(2)\sqrt{2} = 4\sqrt{2}$

6. $\dfrac{25\sqrt{10}}{5\sqrt{5}} = \dfrac{25\sqrt{2}}{5} = 5\sqrt{2}$

7. $\sqrt{24} = \sqrt{4}\sqrt{6} = 2\sqrt{6}$

8. $\sqrt{300} = \sqrt{100}\sqrt{3} = 10\sqrt{3}$

9. $\sqrt{48} = \sqrt{16}\sqrt{3} = 4\sqrt{3}$

10. $(5\sqrt{3})(6\sqrt{5}) = (5)(6)\sqrt{3}\sqrt{5} = 30\sqrt{15}$

11. $(6\sqrt{6})(7\sqrt{2}) = (6)(7)\sqrt{6}\sqrt{2} =$
    $42\sqrt{12} = 42\sqrt{4}\sqrt{3} =$
    $42(2)\sqrt{3} = 84\sqrt{3}$

12. $(2\sqrt{3})(2\sqrt{3}) = (2)(2)\sqrt{3}\sqrt{3} =$
    $4\sqrt{9} = 4(3) = 12$

13. $\sqrt{5} \approx 2.24$

14. $\sqrt{3} \approx 1.73$

15. $\sqrt{14} \approx 3.74$

16. radical

17. numbers; radicals

18. square

Lesson Practice 17B

1. $6\sqrt{7} + 5\sqrt{3} = 6\sqrt{7} + 5\sqrt{3}$:
   cannot be simplified

2. $8\sqrt{3} + 5\sqrt{3} = (8+5)\sqrt{3} = 13\sqrt{3}$

3. $8\sqrt{7} - 7\sqrt{7} = (8-7)\sqrt{7} = 1\sqrt{7} = \sqrt{7}$

4. $13\sqrt{2} + 11\sqrt{2} - 20\sqrt{2} =$
   $(13+11-20)\sqrt{2} = 4\sqrt{2}$

5. $\dfrac{36\sqrt{8}}{6\sqrt{2}} = \dfrac{36\sqrt{4}}{6} = 6\sqrt{4} = 6(2) = 12$

6. $\dfrac{42\sqrt{10}}{7\sqrt{5}} = \dfrac{42\sqrt{2}}{7} = 6\sqrt{2}$

7. $\sqrt{108} = \sqrt{36}\sqrt{3} = 6\sqrt{3}$
   this can also be done
   in smaller steps:
   $\sqrt{108} = \sqrt{9}\sqrt{12} = \sqrt{9}\sqrt{4}\sqrt{3} =$
   $(3)(2)\sqrt{3} = 6\sqrt{3}$

8. $\sqrt{250} = \sqrt{25}\sqrt{10} = 5\sqrt{10}$

9. $\sqrt{180} = \sqrt{36}\sqrt{5} = 6\sqrt{5}$

10. $(6\sqrt{7})(-5\sqrt{7}) = (6)(-5)\sqrt{7}\sqrt{7} =$
    $-30\sqrt{49} = (-30)(7) = -210$

11. $(6\sqrt{2})(4\sqrt{3}) = (6)(4)\sqrt{2}\sqrt{3} = 24\sqrt{6}$

12. $(8)(3\sqrt{20}) = (8)(3)\sqrt{20} = 24\sqrt{20} =$
    $24\sqrt{4}\sqrt{5} = 24(2)\sqrt{5} = 48\sqrt{5}$

13. $\sqrt{6} \approx 2.45$

14. $\sqrt{11} \approx 3.32$

15. $\sqrt{21} \approx 4.58$

16. same

17. $\sqrt{3}\sqrt{3} = \sqrt{9} = 3$

18. whole

Systematic Review 17C

1. $4\sqrt{3} + 5\sqrt{3} = (4+5)\sqrt{3} = 9\sqrt{3}$

2. $\sqrt{7} + 3\sqrt{7} = 1\sqrt{7} + 3\sqrt{7} =$
   $(1+3)\sqrt{7} = 4\sqrt{7}$

3. $16\sqrt{2} - 8\sqrt{2} = (16-8)\sqrt{2} = 8\sqrt{2}$

4. $(\sqrt{7})(\sqrt{10}) = \sqrt{70}$

5. $(\sqrt{5})(2\sqrt{3}) = 2\sqrt{5}\sqrt{3} = 2\sqrt{15}$

6. $\dfrac{\sqrt{28}}{\sqrt{7}} = \sqrt{\dfrac{4}{1}} = \sqrt{4} = 2$

7. $\sqrt{8} = \sqrt{4}\sqrt{2} = 2\sqrt{2}$

8. $\sqrt{28} = \sqrt{4}\sqrt{7} = 2\sqrt{7}$

9. $2\sqrt{2} \approx 2.83$; $2\sqrt{7} \approx 5.29$

10. top:
    V = area of base triangle × height =
    $\dfrac{1}{2}(9)(8) \times 15 = 540$ m^3
    bottom:
    $V = (9)(15)(7) = 945$ m^3
    total:
    $V = 540 + 945 = 1,485$ m^3

11. "roof":
    $SA = (2)(10)(15) = 300$ m^2
    triangles:
    $SA = (2)\dfrac{1}{2}(9)(8) = 72$ m^2
    sides:
    $SA = (2)(7)(15) + (2)(7)(9) =$
    $210 + 126 = 336$ m^2
    bottom:
    $SA = (9)(15) = 135$ m^2
    total:
    $SA = 300 + 72 + 336 + 135 = 843$ m^2

12. $V = \dfrac{1}{3}Bh = \dfrac{1}{3}\pi r^2 h \approx$
    $\dfrac{1}{3}(3.14)(11^2)(14.3) \approx 1,811.05$ ft^3

13. $V = \dfrac{4}{3}\pi r^3 \approx \dfrac{4}{3}(3.14)(2.4^3)$
    ≈ 57.88 in^3

14. $A = bh = (6)(6) = 36$ in^2

15. $A = \pi r^2 \approx (3.14)(3^2) = 28.26$ in^2

16. $A = 36 - 28.26 = 7.74$ in^2

17. TMS

18. vertical

19. alternate interior

20. 70°; vertical angles

Systematic Review 17D

1. $9\sqrt{6} + 23\sqrt{6} = (9+23)\sqrt{6} = 32\sqrt{6}$

2. $7\sqrt{2} + 8\sqrt{2} = (7+8)\sqrt{2} = 15\sqrt{2}$

3. $\sqrt{7} - 5\sqrt{7} = 1\sqrt{7} - 5\sqrt{7} =$
   $(1-5)\sqrt{7} = -4\sqrt{7}$

4. $(3\sqrt{6})(2\sqrt{7}) = (3)(2)\sqrt{6}\sqrt{7} = 6\sqrt{42}$

5. $(\sqrt{11})(\sqrt{11}) = \sqrt{121} = 11$

6. $\dfrac{2\sqrt{30}}{\sqrt{5}} = \dfrac{2\sqrt{6}}{1} = 2\sqrt{6}$

7. $\sqrt{12} = \sqrt{4}\sqrt{3} = 2\sqrt{3}$

8. $\sqrt{200} = \sqrt{100}\sqrt{2} = 10\sqrt{2}$

9. $2\sqrt{3} \approx 3.46$; $10\sqrt{2} \approx 14.14$

10. $V = Bh = \pi r^2 h \approx (3.14)(2^2)(16)$
    $= 200.96 \text{ cm}^3$

11. $SA = 2\pi r^2 + 2\pi rh \approx$
    $(2)(3.14)(2^2) + (2)(3.14)(2)(16) =$
    $25.12 + 200.96 = 226.08 \text{ cm}^2$

12. $A = \dfrac{4+5}{2}(6) = 27 \text{ units}^2$

13. $A = \pi r^2 \approx (3.14)(2.4^2) \oplus 18.09 \text{ in}^2$

14. arc

15. interior angle $= 180° - 45° = 135°$

16. $540° = (N-2)180°$
    $540° = 180N° - 360°$
    $900° = 180N°$
    $\dfrac{900°}{180°} = N = 5$ sides

17. CFD or BFC or AFD: all are 90°

18. 5 cm

19. ADF: line BD is a transversal cutting lines AD and BC, which are parallel, because they are opposite sides of a rhombus. ∠CBF and ∠ADF are alternate interior angles.

20. 30°; alternate interior angles

Systematic Review 17E

1. $5\sqrt{5} + 2\sqrt{2} = 5\sqrt{5} + 2\sqrt{2}$:
   cannot be simplified

2. $3\sqrt{5} + 8\sqrt{5} = (3+8)\sqrt{5} = 11\sqrt{5}$

3. $13\sqrt{10} - 15\sqrt{10} =$
   $(13-15)\sqrt{10} = -2\sqrt{10}$

4. $(5\sqrt{10})(3\sqrt{13}) = (5)(3)\sqrt{10}\sqrt{13}$
   $15\sqrt{130}$

5. $(3\sqrt{8})(2\sqrt{8}) = (3)(2)\sqrt{8}\sqrt{8} =$
   $6\sqrt{64} = 6(8) = 48$

6. $\dfrac{8\sqrt{12}}{2\sqrt{6}} = \dfrac{8\sqrt{2}}{2} = \dfrac{4\sqrt{2}}{1} = 4\sqrt{2}$

7. $\sqrt{27} = \sqrt{9}\sqrt{3} = 3\sqrt{3}$

8. $\sqrt{20} = \sqrt{4}\sqrt{5} = 2\sqrt{5}$

9. $3\sqrt{3} \approx 5.20$; $2\sqrt{5} \approx 4.47$

10. $V = \dfrac{1}{3}Bh = \dfrac{1}{3}(21)(21)(13) =$
    $1,911 \text{ mm}^3$

11. triangles:
    $SA = (4)(\dfrac{1}{2})(21)(15) = 630$
    base:
    $SA = (21)(21) = 441$
    total:
    $SA = 630 + 441 = 1,071 \text{ mm}^2$

12. $A = bh = (11)(8) = 88 \text{ units}^2$

13. $C = \pi d \approx (3.14)(6.4) \approx 20.10 \text{ in}$

14. isosceles

15. interior angle $= 180° - 72° = 108°$

16. $360° \div 72° = 5$ sides

17. $X^2 + 10X + 16 = (X+8)(X+2)$

18. $X^2 + 5X + 6 = (X+2)(X+3)$

19. $X^2 + 8X + 7 = (X+7)(X+1)$

20. $X^2 + 9X + 20 = (X+5)(X+4)$

Lesson Practice 18A

1. right

2. legs

3. hypotenuse

4. Pythagorean

5. the hypotenuse squared;
If the leg squared plus the leg squared equals the hypotenuse squared, the triangle is a right triangle.

6. $8^2 + 9^2 = H^2$
$64 + 81 = H^2$
$145 = H^2$
$\sqrt{145} = H$

7. $5^2 + 5^2 = H^2$
$25 + 25 = H^2$
$50 = H^2$
$\sqrt{50} = H$
$\sqrt{25}\sqrt{2} = H$
$5\sqrt{2} = H$

8. $7^2 + L^2 = 12^2$
$49 + L^2 = 144$
$L^2 = 95$
$L = \sqrt{95}$

9. $12^2 + L^2 = 13^2$
$144 + L^2 = 169$
$L^2 = 25$
$L = \sqrt{25}$
$L = 5$

10. $5^2 + 6^2 = 8^2$
$25 + 36 = 64$
$61 = 64$: false
not a right triangle

11. $2^2 + 4^2 = 6^2$
$4 + 16 = 36$
$20 = 36$: false
not a right triangle

12. $6^2 + 8^2 = 10^2$
$36 + 64 = 100$
$100 = 100$: true
is a right triangle

Lesson Practice 18B

1. 90°
2. legs
3. hypotenuse
4. Pythagorean theorem
5. the hypotenuse squared, then the triangle is a right triangle. If a triangle is a right triangle, then one leg squared plus the other leg squared equals the hypotenuse squared.

6. $7^2 + 10^2 = H^2$
$49 + 100 = H^2$
$149 = H^2$
$\sqrt{149} = H$

7. $3^2 + 3^2 = H^2$
$9 + 9 = H^2$
$18 = H^2$
$\sqrt{18} = H$
$\sqrt{9}\sqrt{2} = H$
$3\sqrt{2} = H$

8. $24^2 + L^2 = 25^2$
$576 + L^2 = 625$
$L^2 = 49$
$L = \sqrt{49}$
$L = 7$

9. $\left(5\sqrt{2}\right)^2 + L^2 = \left(10\sqrt{3}\right)^2$
$(5)(5)\sqrt{2}\sqrt{2} + L^2 = (10)(10)\sqrt{3}\sqrt{3}$
$25\sqrt{4} + L^2 = 100\sqrt{9}$
$25(2) + L^2 = 100(3)$
$50 + L^2 = 300$
$L^2 = 250$
$L = \sqrt{250}$
$L = \sqrt{25}\sqrt{10}$
$L = 5\sqrt{10}$

10. $4^2 + 5^2 = 6^2$
$16 + 25 = 36$
$41 = 36$: false
not a right triangle

11. $10^2 + 24^2 = 26^2$
$100 + 576 = 676$
$676 = 676$: true
is a right triangle

12. $12^2 + 16^2 = 20^2$
$144 + 256 = 400$
$400 = 400$: true
is a right triangle

Systematic Review 18C

1. $7^2 + 10^2 = 49 + 100 = 149$
$\sqrt{149}$ is between 12 and 13
Any answer that is close
is acceptable.

2. $7^2 + 10^2 = H^2$

3. $7^2 + 10^2 = H^2$
$49 + 100 = H^2$
$149 = H^2$
$\sqrt{149} = H$

4. $\sqrt{149} = \sqrt{149}$:
cannot be simplified

5. $4^2 - 3^2 = 16 - 9 = 7$
$\sqrt{7}$ is between 2 and 3

6. $3^2 + L^2 = 4^2$

7. $3^2 + L^2 = 4^2$
$9 + L^2 = 16$
$L^2 = 7$
$L = \sqrt{7}$

8. $\sqrt{7} \approx 2.65$

9. no

10. $5^2 + 7^2 \neq 9^2$

11. hypotenuse

12. $A = \frac{1}{2}bh = \frac{1}{2}(7)(9) = 31.5 \text{ in}^2$

13. $A = \pi r^2 \approx (3.14)(2^2) = 12.56 \text{ in}^2$

14. $A = 31.5 - 12.56 = 18.94 \text{ in}^2$

15. $11\sqrt{5} + 7\sqrt{5} = (11 + 7)\sqrt{5} = 18\sqrt{5}$

16. $(9\sqrt{2})(5\sqrt{2}) = (9)(5)\sqrt{2}\sqrt{2} =$
$45\sqrt{4} = 45(2) = 90$

17. $\frac{9\sqrt{200}}{12\sqrt{2}} = \frac{3\sqrt{100}}{4} = \frac{3(10)}{4} =$
$\frac{30}{4} = 7\frac{1}{2}$ or 7.5

18. $A = \frac{1}{2}bh = \frac{1}{2}(4)(2\sqrt{3}) =$
$(2)(2\sqrt{3}) = 4\sqrt{3} \text{ in}^2$

19. $A = (5)(4\sqrt{3}) = 20\sqrt{3} \text{ in}^2$

20. $P = (5)(4) = 20 \text{ in}$

Systematic Review 18D

1. $7^2 + 7^2 = 49 + 49 = 98$
$\sqrt{98}$ is between 9 and 10

2. $7^2 + 7^2 = B^2$

3. $7^2 + 7^2 = B^2$
$49 + 49 = B^2$
$98 = B^2$
$\sqrt{98} = B$

4. $\sqrt{98} = \sqrt{49}\sqrt{2} = 7\sqrt{2}$

5. $18^2 - (9\sqrt{3})^2 = 324 - 81\sqrt{9} =$
$324 - 81(3) = 324 - 243 = 81$
$\sqrt{81} = 9$

6. $18^2 - (9\sqrt{3})^2 = M^2$

7. $18^2 - (9\sqrt{3})^2 = M^2$
$324 - 81\sqrt{9} = M^2$
$324 - 81(3) = M^2$
$324 - 243 = M^2$
$81 = M^2$
$9 = M$

8. $9\sqrt{3} \approx 15.59$

9. yes

10. $24^2 + 10^2 = 26^2$
    $576 + 100 = 676$
    $676 = 676$: true

11. equilateral

12. yes: they are both radii of the circle

13. $50°$

14. $m\angle OAB + m\angle OBA + m\angle AOB = 180°$
    $m\angle OAB + m\angle OBA + 50° = 180°$
    $m\angle OAB + m\angle OBA = 130°$
    $m\angle OAB = 130° \div 2 = 65°$

15. $\sqrt{3} + \sqrt{3} = 2\sqrt{3}$

16. $\left(\sqrt{6}\right)\left(-\sqrt{24}\right) = -\sqrt{144} = -12$

17. $\dfrac{2\sqrt{15}}{\sqrt{5}} = \dfrac{2\sqrt{3}}{1} = 2\sqrt{3}$

18. $A = \dfrac{1}{2}bh = \dfrac{1}{2}(12)\left(5\sqrt{6}\right) = 30\sqrt{6}$ in^2

    6 triangles

19. $30\sqrt{6}\,(6) = 180\sqrt{6}$ in^2

20. $P = 6(12) = 72$ in

Systematic Review 18E

1. $6^2 + \left(2\sqrt{3}\right)^2 = 36 + 4\left(\sqrt{9}\right) =$
   $36 + 4(3) = 36 + 12 = 48$
   $\sqrt{48}$ is between 6 and 7

2. $R^2 = 6^2 + \left(2\sqrt{3}\right)^2$

3. $R^2 = 6^2 + \left(2\sqrt{3}\right)^2$
   $R^2 = 36 + 4\left(\sqrt{9}\right)$
   $R^2 = 36 + 4(3)$
   $R^2 = 36 + 12$
   $R^2 = 48$
   $R = \sqrt{48}$

4. $\sqrt{48} = \sqrt{16}\sqrt{3} = 4\sqrt{3}$

5. $11^2 + 11^2 = 121 + 121 = 242$
   $\sqrt{242}$ is between 15 and 16

6. $T^2 = 11^2 + 11^2$

7. $T^2 = 11^2 + 11^2$
   $T^2 = 121 + 121$
   $T^2 = 242$
   $T = \sqrt{242} = \sqrt{121}\sqrt{2} = 11\sqrt{2}$

8. $11\sqrt{2} \approx 15.56$

9. yes $\left(2\sqrt{7}\right)^2 + \left(2\sqrt{3}\right)^2 = \left(2\sqrt{10}\right)^2$
    $4\sqrt{49} + 4\sqrt{9} = 4\sqrt{100}$
    $4(7) + 4(3) = 4(10)$
    $28 + 12 = 40$
    $40 = 40$: true

10. The sum of the legs squared equals the hypotenuse squared.

11. yes: It would be a right triangle with equal legs. $(45° - 45° - 90°)$

12. $V = Bh = \pi r^2 h \approx (3.14)(2+3)^2(12)$
    $= 942$ m^3

13. $V = Bh = \pi r^2 h \approx (3.14)(2^2)(12)$
    $= 150.72$ m^3

14. $V = 942 - 150.72 = 791.28$ m^3

15. $5\sqrt{2} - 4\sqrt{3} = 5\sqrt{2} - 4\sqrt{3}$:
    cannot be simplified

16. $\left(5\sqrt{6}\right)\left(2\sqrt{10}\right) =$
    $(5)(2)\sqrt{6}\sqrt{10} = 10\sqrt{60} =$
    $10\sqrt{4}\sqrt{15} = 10(2)\left(\sqrt{15}\right) = 20\sqrt{15}$

17. $\dfrac{\sqrt{100}}{\sqrt{25}} = \dfrac{\sqrt{4}}{1} = \dfrac{2}{1} = 2$ or:
    $\dfrac{\sqrt{100}}{\sqrt{25}} = \dfrac{10}{5} = 2$

18. $X^2 + 3X - 10 = (X + 5)(X - 2)$

19. $X^2 - 2X - 3 = (X - 3)(X + 1)$

20. $X^2 + X - 6 = (X + 3)(X - 2)$

Lesson Practice 19A

1. $\dfrac{9}{\sqrt{2}} = \dfrac{9\sqrt{2}}{\sqrt{2}\sqrt{2}} = \dfrac{9\sqrt{2}}{\sqrt{4}} = \dfrac{9\sqrt{2}}{2}$

2. $\dfrac{10}{\sqrt{5}} = \dfrac{10\sqrt{5}}{\sqrt{5}\sqrt{5}} = \dfrac{10\sqrt{5}}{\sqrt{25}} =$

$\dfrac{10\sqrt{5}}{5} = \dfrac{2\sqrt{5}}{1} = 2\sqrt{5}$

3. $\dfrac{6}{\sqrt{3}} = \dfrac{6\sqrt{3}}{\sqrt{3}\sqrt{3}} = \dfrac{6\sqrt{3}}{\sqrt{9}} =$

$\dfrac{6\sqrt{3}}{3} = \dfrac{2\sqrt{3}}{1} = 2\sqrt{3}$

4. $\dfrac{8}{\sqrt{3}} = \dfrac{8\sqrt{3}}{\sqrt{3}\sqrt{3}} = \dfrac{8\sqrt{3}}{\sqrt{9}} = \dfrac{8\sqrt{3}}{3}$

5. $\dfrac{15}{\sqrt{5}} = \dfrac{15\sqrt{5}}{\sqrt{5}\sqrt{5}} = \dfrac{15\sqrt{5}}{\sqrt{25}} =$

$\dfrac{15\sqrt{5}}{5} = \dfrac{3\sqrt{5}}{1} = 3\sqrt{5}$

6. $\dfrac{9}{\sqrt{3}} = \dfrac{9\sqrt{3}}{\sqrt{3}\sqrt{3}} = \dfrac{9\sqrt{3}}{\sqrt{9}} =$

$\dfrac{9\sqrt{3}}{3} = \dfrac{3\sqrt{3}}{1} = 3\sqrt{3}$

7. $\dfrac{3}{\sqrt{2}} + \dfrac{6}{\sqrt{5}} = \dfrac{3\sqrt{2}}{\sqrt{2}\sqrt{2}} + \dfrac{6\sqrt{5}}{\sqrt{5}\sqrt{5}} =$

$\dfrac{3\sqrt{2}}{\sqrt{4}} + \dfrac{6\sqrt{5}}{\sqrt{25}} = \dfrac{3\sqrt{2}}{2} + \dfrac{6\sqrt{5}}{5} =$

$\dfrac{3\sqrt{2}(5)}{2(5)} + \dfrac{6\sqrt{5}(2)}{5(2)} = \dfrac{15\sqrt{2}}{10} + \dfrac{12\sqrt{5}}{10} =$

$\dfrac{15\sqrt{2}+12\sqrt{5}}{10}$: Note that although 10 and 15 have a common factor, and 10 and 12 have a common factor, there is no factor that is common to all three terms, so this fraction cannot be reduced.

See the next solution for an example of one that can be reduced.

8. $\dfrac{4}{\sqrt{5}} + \dfrac{2}{\sqrt{6}} = \dfrac{4\sqrt{5}}{\sqrt{5}\sqrt{5}} + \dfrac{2\sqrt{6}}{\sqrt{6}\sqrt{6}} =$

$\dfrac{4\sqrt{5}}{\sqrt{25}} + \dfrac{2\sqrt{6}}{\sqrt{36}} = \dfrac{4\sqrt{5}}{5} + \dfrac{2\sqrt{6}}{6} =$

$\dfrac{4\sqrt{5}(6)}{5(6)} + \dfrac{2\sqrt{6}(5)}{6(5)} = \dfrac{24\sqrt{5}}{30} + \dfrac{10\sqrt{6}}{30} =$

$\dfrac{24\sqrt{5}+10\sqrt{6}}{30} = \dfrac{2(12\sqrt{5}+5\sqrt{6})}{2(15)} =$

$\dfrac{12\sqrt{5}+5\sqrt{6}}{15}$

9. $\dfrac{4}{\sqrt{2}} + \dfrac{9}{\sqrt{5}} = \dfrac{4\sqrt{2}}{\sqrt{2}\sqrt{2}} + \dfrac{9\sqrt{5}}{\sqrt{5}\sqrt{5}} =$

$\dfrac{4\sqrt{2}}{\sqrt{4}} + \dfrac{9\sqrt{5}}{\sqrt{25}} = \dfrac{4\sqrt{2}}{2} + \dfrac{9\sqrt{5}}{5} =$

$\dfrac{4\sqrt{2}(5)}{2(5)} + \dfrac{9\sqrt{5}(2)}{5(2)} = \dfrac{20\sqrt{2}}{10} + \dfrac{18\sqrt{5}}{10} =$

$\dfrac{20\sqrt{2}+18\sqrt{5}}{10} = \dfrac{2(10\sqrt{2}+9\sqrt{5})}{2(5)} =$

$\dfrac{10\sqrt{2}+9\sqrt{5}}{5}$

10. $\dfrac{5}{\sqrt{10}} - \dfrac{3}{\sqrt{8}} = \dfrac{5\sqrt{10}}{\sqrt{10}\sqrt{10}} - \dfrac{3\sqrt{8}}{\sqrt{8}\sqrt{8}} =$

$\dfrac{5\sqrt{10}}{\sqrt{100}} - \dfrac{3\sqrt{8}}{\sqrt{64}} = \dfrac{5\sqrt{10}}{10} - \dfrac{3\sqrt{8}}{8} =$

$\dfrac{\sqrt{10}}{2} - \dfrac{3\sqrt{8}}{8} = \dfrac{\sqrt{10}(4)}{2(4)} - \dfrac{3\sqrt{8}}{8} =$

$\dfrac{4\sqrt{10}}{8} - \dfrac{3\sqrt{8}}{8} = \dfrac{4\sqrt{10}-3\sqrt{8}}{8} =$

$\dfrac{4\sqrt{10}-3\sqrt{4}\sqrt{2}}{8} = \dfrac{4\sqrt{10}-3(2)\sqrt{2}}{8} =$

$\dfrac{2(2\sqrt{10}-3\sqrt{2})}{2(4)} = \dfrac{2\sqrt{10}-3\sqrt{2}}{4}$

11. $\dfrac{5}{\sqrt{7}} + \dfrac{6}{\sqrt{2}} = \dfrac{5\sqrt{7}}{\sqrt{7}\sqrt{7}} + \dfrac{6\sqrt{2}}{\sqrt{2}\sqrt{2}} =$

$\dfrac{5\sqrt{7}}{\sqrt{49}} + \dfrac{6\sqrt{2}}{\sqrt{4}} = \dfrac{5\sqrt{7}}{7} + \dfrac{6\sqrt{2}}{2} =$

$\dfrac{5\sqrt{7}(2)}{7(2)} + \dfrac{6\sqrt{2}(7)}{2(7)} =$

$\dfrac{10\sqrt{7}}{14} + \dfrac{42\sqrt{2}}{14} = \dfrac{10\sqrt{7}+42\sqrt{2}}{14} =$

$\dfrac{2(5\sqrt{7}+21\sqrt{2})}{2(7)} = \dfrac{5\sqrt{7}+21\sqrt{2}}{7}$

12. $\dfrac{5\sqrt{2}}{\sqrt{3}} - \dfrac{4\sqrt{3}}{\sqrt{5}} = \dfrac{5\sqrt{2}\sqrt{3}}{\sqrt{3}\sqrt{3}} - \dfrac{4\sqrt{3}\sqrt{5}}{\sqrt{5}\sqrt{5}} =$

$\dfrac{5\sqrt{6}}{\sqrt{9}} - \dfrac{4\sqrt{15}}{\sqrt{25}} = \dfrac{5\sqrt{6}}{3} - \dfrac{4\sqrt{15}}{5} =$

$\dfrac{5\sqrt{6}(5)}{3(5)} - \dfrac{4\sqrt{15}(3)}{5(3)} =$

$\dfrac{25\sqrt{6}}{15} - \dfrac{12\sqrt{15}}{15} = \dfrac{25\sqrt{6}-12\sqrt{15}}{15}$

13. denominator

14. one

15. denominator; common

Lesson Practice 19B

1. $\dfrac{11\sqrt{5}}{\sqrt{5}} = \dfrac{11}{1} = 11$

2. $\dfrac{18}{\sqrt{2}} = \dfrac{18\sqrt{2}}{\sqrt{2}\sqrt{2}} = \dfrac{18\sqrt{2}}{\sqrt{4}} = \dfrac{18\sqrt{2}}{2} =$

   $\dfrac{9\sqrt{2}}{1} = 9\sqrt{2}$

3. $\dfrac{12}{\sqrt{6}} = \dfrac{12\sqrt{6}}{\sqrt{6}\sqrt{6}} = \dfrac{12\sqrt{6}}{\sqrt{36}} = \dfrac{12\sqrt{6}}{6} =$

   $\dfrac{2\sqrt{6}}{1} = 2\sqrt{6}$

4. $\dfrac{7}{\sqrt{2}} = \dfrac{7\sqrt{2}}{\sqrt{2}\sqrt{2}} = \dfrac{7\sqrt{2}}{\sqrt{4}} = \dfrac{7\sqrt{2}}{2}$

5. $\dfrac{6}{\sqrt{3}} = \dfrac{6\sqrt{3}}{\sqrt{3}\sqrt{3}} = \dfrac{6\sqrt{3}}{\sqrt{9}} = \dfrac{6\sqrt{3}}{3} =$

   $\dfrac{2\sqrt{3}}{1} = 2\sqrt{3}$

6. $\dfrac{9\sqrt{6}}{\sqrt{5}} = \dfrac{9\sqrt{6}\sqrt{5}}{\sqrt{5}\sqrt{5}} = \dfrac{9\sqrt{30}}{\sqrt{25}} = \dfrac{9\sqrt{30}}{5}$

7. $\dfrac{4\sqrt{7}}{\sqrt{2}} + \dfrac{3\sqrt{7}}{\sqrt{2}} = \dfrac{4\sqrt{7}+3\sqrt{7}}{\sqrt{2}} = \dfrac{7\sqrt{7}}{\sqrt{2}} =$

   $\dfrac{7\sqrt{7}\sqrt{2}}{\sqrt{2}\sqrt{2}} = \dfrac{7\sqrt{14}}{\sqrt{4}} = \dfrac{7\sqrt{14}}{2}$

8. $\dfrac{\sqrt{12}}{9} - \dfrac{\sqrt{18}}{4} = \dfrac{\sqrt{4}\sqrt{3}}{9} - \dfrac{\sqrt{9}\sqrt{2}}{4} =$

   $\dfrac{2\sqrt{3}}{9} - \dfrac{3\sqrt{2}}{4} = \dfrac{2\sqrt{3}(4)}{9(4)} - \dfrac{3\sqrt{2}(9)}{4(9)} =$

   $\dfrac{8\sqrt{3}}{36} - \dfrac{27\sqrt{2}}{36} = \dfrac{8\sqrt{3}-27\sqrt{2}}{36}$

9. $\dfrac{5}{\sqrt{2}} + \dfrac{7}{\sqrt{8}} = \dfrac{5\sqrt{2}}{\sqrt{2}\sqrt{2}} + \dfrac{7\sqrt{8}}{\sqrt{8}\sqrt{8}} =$

   $\dfrac{5\sqrt{2}}{\sqrt{4}} + \dfrac{7\sqrt{8}}{\sqrt{64}} = \dfrac{5\sqrt{2}}{2} + \dfrac{7\sqrt{8}}{8} =$

   $\dfrac{5\sqrt{2}}{2} + \dfrac{7\sqrt{4}\sqrt{2}}{8} = \dfrac{5\sqrt{2}}{2} + \dfrac{7(2)\sqrt{2}}{8} =$

   $\dfrac{5\sqrt{2}(2)}{2(2)} + \dfrac{7\sqrt{2}}{4} = \dfrac{10\sqrt{2}}{4} + \dfrac{7\sqrt{2}}{4} = \dfrac{17\sqrt{2}}{4}$

10. $\dfrac{8\sqrt{3}}{\sqrt{2}} - \dfrac{5\sqrt{6}}{\sqrt{5}} = \dfrac{8\sqrt{3}\sqrt{2}}{\sqrt{2}\sqrt{2}} - \dfrac{5\sqrt{6}\sqrt{5}}{\sqrt{5}\sqrt{5}} =$

    $\dfrac{8\sqrt{6}}{\sqrt{4}} - \dfrac{5\sqrt{30}}{\sqrt{25}} = \dfrac{8\sqrt{6}}{2} - \dfrac{5\sqrt{30}}{5} =$

    $\dfrac{8\sqrt{6}(5)}{2(5)} - \dfrac{5\sqrt{30}(2)}{5(2)} = \dfrac{40\sqrt{6}}{10} - \dfrac{10\sqrt{30}}{10} =$

    $\dfrac{40\sqrt{6}-10\sqrt{30}}{10} = \dfrac{10(4\sqrt{6}-\sqrt{30})}{10(1)} =$

    $4\sqrt{6}-\sqrt{30}$

11. $\dfrac{3}{\sqrt{5}} + \dfrac{7}{\sqrt{2}} = \dfrac{3\sqrt{5}}{\sqrt{5}\sqrt{5}} + \dfrac{7\sqrt{2}}{\sqrt{2}\sqrt{2}} =$

    $\dfrac{3\sqrt{5}}{\sqrt{25}} + \dfrac{7\sqrt{2}}{\sqrt{4}} = \dfrac{3\sqrt{5}}{5} + \dfrac{7\sqrt{2}}{2} =$

    $\dfrac{3\sqrt{5}(2)}{5(2)} + \dfrac{7\sqrt{2}(5)}{2(5)} = \dfrac{6\sqrt{5}}{10} + \dfrac{35\sqrt{2}}{10} =$

    $\dfrac{6\sqrt{5}+35\sqrt{2}}{10}$

12. $\dfrac{4\sqrt{11}}{\sqrt{3}} + \dfrac{2\sqrt{5}}{\sqrt{7}} = \dfrac{4\sqrt{11}\sqrt{3}}{\sqrt{3}\sqrt{3}} + \dfrac{2\sqrt{5}\sqrt{7}}{\sqrt{7}\sqrt{7}} =$

    $\dfrac{4\sqrt{33}}{\sqrt{9}} + \dfrac{2\sqrt{35}}{\sqrt{49}} = \dfrac{4\sqrt{33}}{3} + \dfrac{2\sqrt{35}}{7} =$

    $\dfrac{4\sqrt{33}(7)}{3(7)} + \dfrac{2\sqrt{35}(3)}{7(3)} = \dfrac{28\sqrt{33}}{21} + \dfrac{6\sqrt{35}}{21} =$

    $\dfrac{28\sqrt{33}+6\sqrt{35}}{21}$

13. one

14. radical or square root

15. common denominator

Systematic Review 19C

1. $\dfrac{6}{\sqrt{7}} = \dfrac{6\sqrt{7}}{\sqrt{7}\sqrt{7}} = \dfrac{6\sqrt{7}}{\sqrt{49}} = \dfrac{6\sqrt{7}}{7}$

2. $\dfrac{8}{\sqrt{2}} = \dfrac{8\sqrt{2}}{\sqrt{2}\sqrt{2}} = \dfrac{8\sqrt{2}}{\sqrt{4}} = \dfrac{8\sqrt{2}}{2} =$

   $\dfrac{4\sqrt{2}}{1} = 4\sqrt{2}$

3. $\dfrac{6\sqrt{2}}{\sqrt{3}} = \dfrac{6\sqrt{2}\sqrt{3}}{\sqrt{3}\sqrt{3}} = \dfrac{6\sqrt{6}}{\sqrt{9}} = \dfrac{6\sqrt{6}}{3} = 2\sqrt{6}$

4. $2\sqrt{3} + 6\sqrt{2} = 2\sqrt{3} + 6\sqrt{2}$

   cannot be simplified

5. $\left(4\sqrt{3}\right)\left(7\sqrt{15}\right) =$
$(4)(7)\sqrt{3}\sqrt{15} = 28\sqrt{45} =$
$28\sqrt{9}\sqrt{5} = 28(3)\sqrt{5} = 84\sqrt{5}$

6. $\dfrac{\sqrt{36}}{\sqrt{6}} = \dfrac{\sqrt{6}}{1} = \sqrt{6}$

7. $\dfrac{10\sqrt{10}}{\sqrt{7}} - \dfrac{2\sqrt{6}}{\sqrt{11}} =$
$\dfrac{10\sqrt{10}\sqrt{7}}{\sqrt{7}\sqrt{7}} - \dfrac{2\sqrt{6}\sqrt{11}}{\sqrt{11}\sqrt{11}} =$
$\dfrac{10\sqrt{70}}{\sqrt{49}} - \dfrac{2\sqrt{66}}{\sqrt{121}} = \dfrac{10\sqrt{70}}{7} - \dfrac{2\sqrt{66}}{11} =$
$\dfrac{10\sqrt{70}\,(11)}{7(11)} - \dfrac{2\sqrt{66}\,(7)}{11(7)} =$
$\dfrac{110\sqrt{70}}{77} - \dfrac{14\sqrt{66}}{77} =$
$\dfrac{110\sqrt{70} - 14\sqrt{66}}{77}$

8. $\dfrac{24\sqrt{13}}{\sqrt{3}} + \dfrac{3\sqrt{2}}{\sqrt{3}} = \dfrac{24\sqrt{13}\sqrt{3}}{\sqrt{3}\sqrt{3}} + \dfrac{3\sqrt{2}\sqrt{3}}{\sqrt{3}\sqrt{3}} =$
$\dfrac{24\sqrt{39}}{\sqrt{9}} + \dfrac{3\sqrt{6}}{\sqrt{9}} = \dfrac{24\sqrt{39}}{3} + \dfrac{3\sqrt{6}}{3} =$
$\dfrac{24\sqrt{39} + 3\sqrt{6}}{3} = \dfrac{3\left(8\sqrt{39} + \sqrt{6}\right)}{3(1)} =$
$8\sqrt{39} + \sqrt{6}$

9. $5^2 + 10^2 = 25 + 100 = 125$
$\sqrt{125}$ is between 11 and 12

10. $5^2 + 10^2 = Q^2$

11. $5^2 + 10^2 = Q^2$
$25 + 100 = Q^2$
$125 = Q^2$
$\sqrt{125} = Q = \sqrt{25}\sqrt{5} = 5\sqrt{5}$

12. guess about 12

13. $X^2 + 16^2 = 20^2$
$X^2 + 256 = 400$
$X^2 = 144$

14. $X^2 + 16^2 = 20^2$
$X^2 + 256 = 400$
$X^2 = 144$
$X = 12$

15. $V = \dfrac{1}{3}Bh = \dfrac{1}{3}\pi r^2 h \approx$
$\dfrac{1}{3}(3.14)\left(11^2\right)(14.3) = 1{,}811.05 \text{ in}^3$

16. check with ruler and protractor

17. The measures of the angles of a quadrilateral add up to 360°. In a rhombus, opposite angles are congruent, because they are formed by transversals cutting parallel lines. If two of the angles measure 60°, then the other two would be:
$360° - \left(2 \times 60°\right) =$
$360° - 120° = 240°$
If they add up to 240°, and are equal, then each must have a measure of:
$240° \div 2 = 120°$

18. $A = \dfrac{1}{2}bh = \dfrac{1}{2}\left(5\right)\left(3\sqrt{2}\right) = 7.5\sqrt{2} \text{ in}^2$
8 triangles

19. $(8)\left(7.5\sqrt{2}\right) = 60\sqrt{2} \text{ in}^2$

20. $P = 8(5) = 40 \text{ in}$

Systematic Review 19D

1. $\dfrac{9}{\sqrt{5}} = \dfrac{9\sqrt{5}}{\sqrt{5}\sqrt{5}} = \dfrac{9\sqrt{5}}{\sqrt{25}} = \dfrac{9\sqrt{5}}{5}$

2. $\dfrac{6}{\sqrt{2}} = \dfrac{6\sqrt{2}}{\sqrt{2}\sqrt{2}} = \dfrac{6\sqrt{2}}{\sqrt{4}} = \dfrac{6\sqrt{2}}{2} = 3\sqrt{2}$

3. $\dfrac{5\sqrt{10}}{\sqrt{6}} = \dfrac{5\sqrt{10}\sqrt{6}}{\sqrt{6}\sqrt{6}} = \dfrac{5\sqrt{60}}{\sqrt{36}} = \dfrac{5\sqrt{60}}{6} =$
$\dfrac{5\sqrt{4}\sqrt{15}}{6} = \dfrac{5(2)\sqrt{15}}{6} = \dfrac{5\sqrt{15}}{3}$

4. $5\sqrt{6} + 2\sqrt{10} = 5\sqrt{6} + 2\sqrt{10}:$
cannot be simplified

5. $\left(3\sqrt{8}\right)\left(2\sqrt{5}\right) = (3)(2)\sqrt{8}\sqrt{5} = 6\sqrt{40} =$
$6\sqrt{4}\sqrt{10} = 6(2)\sqrt{10} = 12\sqrt{10}$

6. $\dfrac{2\sqrt{14}}{\sqrt{7}} = \dfrac{2\sqrt{2}}{1} = 2\sqrt{2}$

7. $\dfrac{2\sqrt{5}}{\sqrt{6}} + \dfrac{2\sqrt{2}}{\sqrt{3}} = \dfrac{2\sqrt{5}\sqrt{6}}{\sqrt{6}\sqrt{6}} + \dfrac{2\sqrt{2}\sqrt{3}}{\sqrt{3}\sqrt{3}} =$

$\dfrac{2\sqrt{30}}{\sqrt{36}} + \dfrac{2\sqrt{6}}{\sqrt{9}} = \dfrac{2\sqrt{30}}{6} + \dfrac{2\sqrt{6}}{3} =$

$\dfrac{\sqrt{30}}{3} + \dfrac{2\sqrt{6}}{3} = \dfrac{\sqrt{30}+2\sqrt{6}}{3}$

8. $\dfrac{5\sqrt{11}}{\sqrt{2}} - \dfrac{3\sqrt{5}}{\sqrt{2}} = \dfrac{5\sqrt{11}\sqrt{2}}{\sqrt{2}\sqrt{2}} - \dfrac{3\sqrt{5}\sqrt{2}}{\sqrt{2}\sqrt{2}} =$

$\dfrac{5\sqrt{22}}{\sqrt{4}} - \dfrac{3\sqrt{10}}{\sqrt{4}} = \dfrac{5\sqrt{22}}{2} - \dfrac{3\sqrt{10}}{2} =$

$\dfrac{5\sqrt{22}-3\sqrt{10}}{2}$

9. $9^2 + 11^2 = X^2$

10. $9^2 + 11^2 = X^2$

$81 + 121 = X^2$

$202 = X^2$

$\sqrt{202} = X \oplus 14.21$ units

11. $A = \dfrac{1}{2}bh = \dfrac{1}{2}(9)(11) = 49.5$ units^2

12. $L^2 + 5^2 = 13^2$

13. $L^2 + 5^2 = 13^2$

$L^2 + 25 = 169$

$L^2 = 144$

$L = 12$ units

14. $A = \dfrac{1}{2}bh = \dfrac{1}{2}(5)(12) = 30$ units^2

15. check with protractor

16. 45°: check with protractor

17. $A = \dfrac{1}{2}(7) \times \dfrac{1}{2}(5) \times \pi \approx$

$(3.5)(2.5)(3.14) \approx 27.48$ m^2

18. $A = \pi r^2 \approx (3.14)(3)^2 = 28.26$ in^2

19. $\dfrac{90°}{360°} = \dfrac{1}{4}$

20. $A = \dfrac{1}{4}(28.26) = (.25)(28.26) \approx 7.07$ in^2

Systematic Review 19E

1. $\dfrac{2\sqrt{6}}{2\sqrt{5}} = \dfrac{\sqrt{6}}{\sqrt{5}} = \dfrac{\sqrt{6}\sqrt{5}}{\sqrt{5}\sqrt{5}} = \dfrac{\sqrt{30}}{\sqrt{25}} = \dfrac{\sqrt{30}}{5}$

2. $\dfrac{\sqrt{5}}{\sqrt{10}} = \dfrac{1}{\sqrt{2}} = \dfrac{\sqrt{2}}{\sqrt{2}\sqrt{2}} = \dfrac{\sqrt{2}}{\sqrt{4}} = \dfrac{\sqrt{2}}{2}$

3. $\dfrac{12\sqrt{13}}{2\sqrt{13}} = \dfrac{12}{2} = 6$

4. $8\sqrt{2} + 2\sqrt{14} = 8\sqrt{2} + 2\sqrt{14}$:

cannot be simplified

5. $(2\sqrt{7})(5\sqrt{8}) = (2)(5)\sqrt{7}\sqrt{8} =$

$10\sqrt{56} = 10\sqrt{4}\sqrt{14} =$

$10(2)\sqrt{14} = 20\sqrt{14}$

6. $\dfrac{14}{\sqrt{7}} = \dfrac{14\sqrt{7}}{\sqrt{7}\sqrt{7}} = \dfrac{14\sqrt{7}}{\sqrt{49}} = \dfrac{14\sqrt{7}}{7} =$

$\dfrac{2\sqrt{7}}{1} = 2\sqrt{7}$

7. $\dfrac{15}{\sqrt{5}} + \dfrac{20}{\sqrt{2}} = \dfrac{15\sqrt{5}}{\sqrt{5}\sqrt{5}} + \dfrac{20\sqrt{2}}{\sqrt{2}\sqrt{2}} =$

$\dfrac{15\sqrt{5}}{\sqrt{25}} + \dfrac{20\sqrt{2}}{\sqrt{4}} = \dfrac{15\sqrt{5}}{5} + \dfrac{20\sqrt{2}}{2} =$

$3\sqrt{5} + 10\sqrt{2}$

8. $\dfrac{5\sqrt{2}}{\sqrt{14}} - \dfrac{3\sqrt{2}}{\sqrt{14}} = \dfrac{5\sqrt{2}-3\sqrt{2}}{\sqrt{14}} = \dfrac{2\sqrt{2}}{\sqrt{14}} =$

$\dfrac{2\sqrt{2}\sqrt{14}}{\sqrt{14}\sqrt{14}} = \dfrac{2\sqrt{28}}{\sqrt{196}} = \dfrac{2\sqrt{28}}{14} =$

$\dfrac{\sqrt{28}}{7} = \dfrac{\sqrt{4}\sqrt{7}}{7} = \dfrac{2\sqrt{7}}{7}$

9. $5^2 + 6^2 = X^2$

10. $5^2 + 6^2 = X^2$

$25 + 36 = X^2$

$61 = X^2$

$\sqrt{61} = X^2 \oplus 7.81$ units

11. $A = \dfrac{1}{2}(5)(6) = 15$ units^2

12. $B^2 + (2B)^2 = X^2$

13. $B^2 + (2B)^2 = X^2$

$B^2 + (2B)(2B) = X^2$

$B^2 + 4B^2 = X^2$

$5B^2 = X^2$

$\sqrt{5B^2} = X$

$X = \sqrt{B^2}\sqrt{5} = B\sqrt{5}$ units

14. $A = \frac{1}{2}bh = \frac{1}{2}(B)(2B) = B^2$ units^2

15. $A = \pi r^2 \approx (3.14)(4^2) = 50.24$ in^2

16. sector is $\frac{45°}{360°} = \frac{1}{8}$ of the circle

17. $A = \frac{1}{8}(50.24)$

$= (.125)(50.24) \approx 6.28$ in^2

18. $9^{\frac{1}{2}} = \sqrt{9} = 3$

19. $1,000^{\frac{2}{3}} = \sqrt[3]{1,000}^2 = 10^2 = 100$

20. $4^{\frac{3}{2}} = \left(\sqrt{4}\right)^3 = 2^3 = 8$

Lesson Practice 20A

1. 90°; 45°
2. isosceles
3. hypotenuse
4. equal
5. Pythagorean
6. $\sqrt{2}$
7. $7\sqrt{2}$
8. 7
9. $5\sqrt{2}$
10. 5
11. $3\sqrt{2}$
12. $3\sqrt{2}\sqrt{2} = 3\sqrt{4} = 3(2) = 6$
13. $\frac{8\sqrt{2}}{\sqrt{2}} = \frac{8}{1} = 8$
14. 8

15. Since ABCD is a square, all four sides are equal in length. Therefore, △ABD has two congruent sides, and is a 45° – 45° – 90° triangle.

$10\sqrt{2}$

Lesson Practice 20B

1. false
2. false
3. true
4. true
5. false
6. true
7. $8\sqrt{2}$
8. 8
9. $6\sqrt{3}\sqrt{2} = 6\sqrt{6}$
10. $6\sqrt{3}$
11. $\frac{5\sqrt{2}}{\sqrt{2}} = \frac{5}{1} = 5$
12. 5
13. $\frac{10}{\sqrt{2}} = \frac{10\sqrt{2}}{\sqrt{2}\sqrt{2}} = \frac{10\sqrt{2}}{\sqrt{4}} =$

$\frac{10\sqrt{2}}{2} = \frac{5\sqrt{2}}{1} = 5\sqrt{2}$

14. $5\sqrt{2}$

15. $\frac{14}{\sqrt{2}} = \frac{14\sqrt{2}}{\sqrt{2}\sqrt{2}} = \frac{14\sqrt{2}}{\sqrt{4}} = \frac{14\sqrt{2}}{2} =$

$\frac{7\sqrt{2}}{1} = 7\sqrt{2}$ cm

Systematic Review 20C

1. $5\sqrt{2}$
2. 5
3. $\frac{20\sqrt{2}}{\sqrt{2}} = \frac{20}{1} = 20$
4. 20

5. $H^2 = 3^2 + 7^2$

   $H^2 = 9 + 49$

   $H^2 = 58$

   $H = \sqrt{58}$

6. $\left(11\sqrt{40}\right)\left(9\sqrt{5}\right) = (11)(9)\sqrt{40}\sqrt{5}$

   $= 99\sqrt{200} = 99\sqrt{100}\sqrt{2} =$

   $99(10)\sqrt{2} = 990\sqrt{2}$

7. $\left(2\sqrt{6}\right)\left(10\sqrt{8}\right) = (2)(10)\sqrt{6}\sqrt{8} =$

   $20\sqrt{48} = 20\sqrt{16}\sqrt{3} =$

   $20(4)\sqrt{3} = 80\sqrt{3}$

8. $\dfrac{5\sqrt{6}}{\sqrt{7}} - \dfrac{\sqrt{5}}{\sqrt{6}} = \dfrac{5\sqrt{6}\sqrt{7}}{\sqrt{7}\sqrt{7}} - \dfrac{\sqrt{5}\sqrt{6}}{\sqrt{6}\sqrt{6}} =$

   $\dfrac{5\sqrt{42}}{\sqrt{49}} - \dfrac{\sqrt{30}}{\sqrt{36}} = \dfrac{5\sqrt{42}}{7} - \dfrac{\sqrt{30}}{6} =$

   $\dfrac{5\sqrt{42}(6)}{7(6)} - \dfrac{\sqrt{30}(7)}{6(7)} =$

   $\dfrac{30\sqrt{42}}{42} - \dfrac{7\sqrt{30}}{42} = \dfrac{30\sqrt{42} - 7\sqrt{30}}{42}$

9. check with ruler and protractor

10. $360° - (2 \times 123°) =$

    $360° - 246° = 114°$

    $114° \div 2 = 57°$

11. Figure is a parallelogram

    $A = bh = (6)(3.5) = 21 \text{ cm}^2$

12. $(N - 2)180° \Rightarrow ((5) - 2)180° =$

    $(3)(180°) = 540°$ total

    $540° \div 5$ sides $= 108°$ per side

13. $A = \pi r^2 \approx (3.14)\left(62^2\right)$

    $= 12{,}070.16 \text{ mm}^2$

14. $C = 2\pi r \approx (2)(3.14)(62)$

    $= 389.36 \text{ mm}$

15. $V = \dfrac{4}{3}\pi r^3 \approx \dfrac{4}{3}(3.14)\left(1^3\right) \approx 4.19 \text{ m}^3$

16. $\dfrac{120°}{360°} = \dfrac{1}{3}$

17. circle:

    $A = \pi r^2 \approx (3.14)\left(3^2\right) = 28.26 \text{ units}^2$

    sector: $A = \dfrac{1}{3}(28.26) = 9.42 \text{ units}^2$

18. $A = \dfrac{1}{2}bh = \dfrac{1}{2}(8)\left(4\sqrt{7}\right) = 16\sqrt{7} \text{ in}^2$

    5 triangles

19. $5\left(16\sqrt{7}\right) = 80\sqrt{7} \text{ in}^2$

20. $P = 5(8) = 40 \text{ in}$

Systematic Review 20D

1. $7\sqrt{2}\sqrt{2} = 7\sqrt{4} = 7(2) = 14$

2. $7\sqrt{2}$

3. $\dfrac{25}{\sqrt{2}} = \dfrac{25\sqrt{2}}{\sqrt{2}\sqrt{2}} = \dfrac{25\sqrt{2}}{\sqrt{4}} = \dfrac{25\sqrt{2}}{2}$

4. $\dfrac{25\sqrt{2}}{2}$

5. $H^2 = 6^2 + 8^2$

   $H^2 = 36 + 64$

   $H^2 = 100$

   $H = 10$

6. $\left(12\sqrt{3}\right)\left(4\sqrt{18}\right) = (12)(4)\sqrt{3}\sqrt{18} =$

   $48\sqrt{54} = 48\sqrt{9}\sqrt{6} = 48(3)\sqrt{6} =$

   $144\sqrt{6}$

7. $\left(4\sqrt{5}\right)\left(20\sqrt{2}\right) = (4)(20)\sqrt{5}\sqrt{2} =$

   $80\sqrt{10}$

8. $\dfrac{2\sqrt{25}}{\sqrt{5}} + \dfrac{\sqrt{5}}{\sqrt{25}} = \dfrac{2(5)}{\sqrt{5}} + \dfrac{1}{\sqrt{5}} =$

   $\dfrac{10}{\sqrt{5}} + \dfrac{1}{\sqrt{5}} = \dfrac{11}{\sqrt{5}} =$

   $\dfrac{11\sqrt{5}}{\sqrt{5}\sqrt{5}} = \dfrac{11\sqrt{5}}{\sqrt{25}} = \dfrac{11\sqrt{5}}{5}$

9. Check with ruler and protractor: second pair of sides should

   be $\dfrac{7}{8}$ in. Angles should be 90°.

10. $P = 2\left(1\dfrac{3}{4} + \dfrac{7}{8}\right) = 2\left(\dfrac{7}{4} + \dfrac{7}{8}\right) =$

    $2\left(\dfrac{14}{8} + \dfrac{7}{8}\right) = 2\left(\dfrac{21}{8}\right) =$

    $\dfrac{42}{8} = \dfrac{21}{4} = 5\dfrac{1}{4} \text{ in}$

11. $V = \dfrac{4}{3}\pi r^3$

12. $(N-2)180° \Rightarrow ((8)-2)180° =$
$(6)180° = 1,080°$ total;
$1,080° \div 8 = 135°$ per angle

13. $V = \frac{1}{3}Bh = \frac{1}{3}\pi r^2 h \approx$
$\frac{1}{3}(3.14)(8.3)^2(12.4) = 894.1$ cm^3

14. diameter

15. $V = Bh = \pi r^2 h \approx$
$(3.14)(8^2)(12) = 2,411.52$ cm^3

16. $SA = 2\pi r^2 + 2\pi rh \approx$
$(2)(3.14)(8^2)+(2)(3.14)(8)(12) =$
$401.92 + 602.88 = 1,004.8$ cm^2

17. $\frac{60°}{360°} = \frac{1}{6}$

18. circle:
$A = \pi r^2 \approx (3.14)(4^2) = 50.24$ cm^2
sector: $A = \frac{1}{6}(50.24) \approx 8.37$ cm^2

19. semicircle:
$A = \frac{1}{2}\pi r^2 \approx \frac{1}{2}(3.14)(3^2) = 14.13$ in^2
rectangle:
$A = bh = (9)(6) = 54$ in^2
total:
$A = 14.13 + 54 = 68.13$ in^2

20. perimeter of semicircle is half the
circumference of the circle:
$P = \frac{1}{2}(2\pi r) \approx \frac{1}{2}(2)(3.14)(3) = 9.42$ in
rectangle (exterior lines only):
$P = 9 + 6 + 9 = 24$ in
total: $9.42 + 24 = 33.42$ in

Systematic Review 20E

1. $14\sqrt{3}\sqrt{2} = 14\sqrt{6}$

2. $14\sqrt{3}$

3. $\frac{5\sqrt{2}}{\sqrt{2}} = \frac{5}{1} = 5$

4. 5

5. $H^2 = 4^2 + 10^2$
$H^2 = 16 + 100$
$H^2 = 116$
$H = \sqrt{116} = \sqrt{4}\sqrt{29} = 2\sqrt{29}$

6. $(10\sqrt{2})(3\sqrt{8}) = (10)(3)\sqrt{2}\sqrt{8} =$
$30\sqrt{16} = 30(4) = 120$

7. $(4\sqrt{7})(6\sqrt{3}) = (4)(6)\sqrt{7}\sqrt{3} = 24\sqrt{21}$

8. $\frac{2\sqrt{16}}{\sqrt{2}} + \frac{\sqrt{2}}{\sqrt{16}} = \frac{2(4)}{\sqrt{2}} + \frac{\sqrt{2}}{4} =$
$\frac{8}{\sqrt{2}} + \frac{\sqrt{2}}{4} = \frac{8\sqrt{2}}{\sqrt{2}\sqrt{2}} + \frac{\sqrt{2}}{4} =$
$\frac{8\sqrt{2}}{\sqrt{4}} + \frac{\sqrt{2}}{4} = \frac{8\sqrt{2}}{2} + \frac{\sqrt{2}}{4} =$
$\frac{8\sqrt{2}(2)}{2(2)} + \frac{\sqrt{2}}{4} = \frac{16\sqrt{2}}{4} + \frac{\sqrt{2}}{4} = \frac{17\sqrt{2}}{4}$

9. Use your compass to draw a
circle. Draw the radius, and use
your protractor to measure out
an angle of 210°. See the end
of lesson 4 in the instruction
manual for hints on drawing
or measuring obtuse angles.

10. prism

11. $A = \frac{10+12}{2}(13)$
$= (11)(13) = 143$ units^2

12. $(N-2)180° \Rightarrow ((10)-2)180° =$
$(8)180° = 1,440°$ total;
$1,440° \div 10 = 144°$ per side

13. line or segment or ray

14. angles are congruent

15. $C = 2\pi r \approx (2)(3.14)(6) = 37.68$ in

16. Measure of minor arc BC =
$m\angle BOC = 90°$
$\frac{90°}{360°} = \frac{1}{4}$ so arc is $\frac{1}{4}$ of circle
length of arc = $\frac{1}{4}(37.68) = 9.42$ in

17. $37.68 - 9.42 = 28.26$ in

18. $7X^2 + 28X + 28 =$
$(7)(X^2 + 4X + 4) =$
$(7)(X+2)(X+2)$

19. $3X^2 + 15X - 18 =$
    $(3)(X^2 + 5X - 6) =$
    $(3)(X + 6)(X - 1)$

20. $2X^2 + 11X + 5 =$
    $(2X + 1)(X + 5)$

Lesson Practice 21A

1. 60; 90
2. scalene
3. 2
4. $\sqrt{3}$
5. Pythagorean
6. 2; multiply; $\sqrt{3}$
7. $5\sqrt{3}$
8. $5(2) = 10$
9. $\dfrac{6\sqrt{3}}{\sqrt{3}} = \dfrac{6}{1} = 6$
10. $6(2) = 12$
11. $\dfrac{10\sqrt{3}}{2} = \dfrac{5\sqrt{3}}{1} = 5\sqrt{3}$
12. $5\sqrt{3}\sqrt{3} = 5\sqrt{9} = 5(3) = 15$
13. $8\sqrt{3}$
14. $8(2) = 16$
15. Because of the relationship between the lengths of the sides, we know that $\triangle DBC$ is a 30°–60°–90° triangle, so the measures are 30° and 60° respectively.

Lesson Practice 21B

1. false
2. false
3. true
4. false
5. true
6. true
7. $\dfrac{7\sqrt{3}}{\sqrt{3}} = \dfrac{7}{1} = 7$

8. $7(2) = 14$
9. $\dfrac{8\sqrt{3}}{2} = \dfrac{4\sqrt{3}}{1} = 4\sqrt{3}$
10. $4\sqrt{3}\sqrt{3} = 4\sqrt{9} = 4(3) = 12$
11. $11(2) = 22$
12. $11\sqrt{3}$
13. $\dfrac{12\sqrt{3}}{\sqrt{3}} = \dfrac{12}{1} = 12$
14. $12(2) = 24$
15. radius of circle =
    hypotenuse of triangle =
    $5(2) = 10$ cm

Systematic Review 21C

1. $7(2) = 14$
2. $7\sqrt{3}$
3. $\dfrac{9\sqrt{3}}{\sqrt{3}}(2) = \dfrac{9}{1}(2) = 9(2) = 18$
4. $\dfrac{9\sqrt{3}}{\sqrt{3}} = \dfrac{9}{1} = 9$
5. $2\sqrt{2}\sqrt{2} = 2\sqrt{4} = 2(2) = 4$
6. $2\sqrt{2}$
7. $6\sqrt{2}$
8. 6
9. $(10\sqrt{20})(2\sqrt{7}) = (10)(2)\sqrt{20}\sqrt{7} =$
   $20\sqrt{140} = 20\sqrt{4}\sqrt{35} =$
   $20(2)\sqrt{35} = 40\sqrt{35}$
10. $3\sqrt{5} + 4\sqrt{5} = (3 + 4)\sqrt{5} = 7\sqrt{5}$
11. $\dfrac{4\sqrt{8}}{\sqrt{4}} + \dfrac{\sqrt{5}}{\sqrt{16}} = \dfrac{4\sqrt{8}}{2} + \dfrac{\sqrt{5}}{4} =$
    $\dfrac{4\sqrt{8}(2)}{2(2)} + \dfrac{\sqrt{5}}{4} = \dfrac{8\sqrt{8}}{4} + \dfrac{\sqrt{5}}{4} =$
    $\dfrac{8\sqrt{8} + \sqrt{5}}{4} = \dfrac{8\sqrt{4}\sqrt{2} + \sqrt{5}}{4} =$
    $\dfrac{8(2)\sqrt{2} + \sqrt{5}}{4} = \dfrac{16\sqrt{2} + \sqrt{5}}{4}$

12. $H^2 = 22^2 + 16^2$

$H^2 = 484 + 256$

$H^2 = 740$

$H = \sqrt{740} = \sqrt{4}\sqrt{185} = 2\sqrt{185}$

13. $SA = 4\pi r^2 \approx (4)(3.14)(1.5^2) =$

28.26 in^2

14. see drawing

15. exterior angle $= 180° - 140° = 40°$

$360° \div 40° = 9$ sides

16. $C = 2\pi r \approx (2)(3.14)(7) = 43.96$ in

17. $\dfrac{180°}{360°} = \dfrac{1}{2}$

18. $\dfrac{1}{2}(43.96) = 21.98$ in

19. area of semicircle is half area of circle:

$A = \dfrac{1}{2}\pi r^2 \approx \dfrac{1}{2}(3.14)(2.5^2)$

$\approx 9.81 \text{ cm}^2$

top base of trapezoid is twice the radius:

$A = \dfrac{5+8}{2}(3) = 19.5 \text{ cm}^2$

total:

$9.81 + 19.5 = 29.31 \text{ cm}^2$

20. semicircle:

$C = \dfrac{1}{2}(2\pi r) \approx \dfrac{1}{2}(2)(3.14)(2.5)$

$= 7.85$ cm

trapezoid:

$P = 4.5 + 8 + 4 = 16.5$ cm

total:

$7.85 + 16.5 = 24.35$ cm

Systematic Review 21D

1. $\dfrac{4\sqrt{2}}{\sqrt{3}} = \dfrac{4\sqrt{2}\sqrt{3}}{\sqrt{3}\sqrt{3}} = \dfrac{4\sqrt{6}}{\sqrt{9}} = \dfrac{4\sqrt{6}}{3}$

2. $\dfrac{4\sqrt{6}}{3}(2) = \dfrac{8\sqrt{6}}{3}$

3. $\dfrac{12\sqrt{3}}{2} = \dfrac{6\sqrt{3}}{1} = 6\sqrt{3}$

4. $\dfrac{6\sqrt{3}}{\sqrt{3}} = \dfrac{6}{1} = 6$

5. $15\sqrt{2}$

6. 15

7. $5\sqrt{2}\sqrt{2} = 5\sqrt{4} = 5(2) = 10$

8. $5\sqrt{2}$

9. $(4\sqrt{4})(2\sqrt{4}) = (4)(2)\sqrt{4}\sqrt{4} =$

$8\sqrt{16} = 8(4) = 32$

10. $2\sqrt{18} + 5\sqrt{9} = 2\sqrt{9}\sqrt{2} + 5(3) =$

$2(3)\sqrt{2} + 15 = 6\sqrt{2} + 15$

11. $\dfrac{3\sqrt{6}}{\sqrt{7}} + \dfrac{4\sqrt{6}}{\sqrt{5}} = \dfrac{3\sqrt{6}\sqrt{7}}{\sqrt{7}\sqrt{7}} + \dfrac{4\sqrt{6}\sqrt{5}}{\sqrt{5}\sqrt{5}} =$

$\dfrac{3\sqrt{42}}{\sqrt{49}} + \dfrac{4\sqrt{30}}{\sqrt{25}} = \dfrac{3\sqrt{42}}{7} + \dfrac{4\sqrt{30}}{5} =$

$\dfrac{3\sqrt{42}(5)}{7(5)} + \dfrac{4\sqrt{30}(7)}{5(7)} =$

$\dfrac{15\sqrt{42}}{35} + \dfrac{28\sqrt{30}}{35} =$

$\dfrac{15\sqrt{42} + 28\sqrt{30}}{35}$

12. $13^2 + L^2 = 17^2$

$169 + L^2 = 289$

$L^2 = 120$

$L = \sqrt{120} = \sqrt{4}\sqrt{30} = 2\sqrt{30}$

13. $SA = 4\pi r^2 \approx (4)(3.14)(2)^2$

$= 50.24 \text{ in}^2$

14. see drawing

15. exterior angle $= 180° - 150° = 30°$

$360° \div 30° = 12$ sides

16. $C = 2\pi r \approx (2)(3.14)(4) = 25.12$ in

17. $\dfrac{270°}{360°} = \dfrac{3}{4}$

18. $\frac{3}{4}(25.12) = (.75)(25.12) = 18.84$ in

19. V = Area of base times height =
    $\frac{1}{2}(6.2)(9.1)(13) = 366.73$ in^3

20. triangles:
    $A = (2)\frac{1}{2}bh = (2)\frac{1}{2}(6.2)(9.1)$
    $= 56.42$ in^2
    large rectangles:
    $A = (2)(11)(13) = 286$ in^2
    small rectangle:
    $A = (6.2)(13) = 80.6$ in^2
    total:
    $A = 56.42 + 286 + 80.6$
    $= 423.02$ in^2

Systematic Review 21E

1. $\frac{\sqrt{3}}{\sqrt{3}} = 1$

2. $1(2) = 2$

3. $\frac{16\sqrt{7}\sqrt{3}}{2} = \frac{16\sqrt{21}}{2} = \frac{8\sqrt{21}}{1} = 8\sqrt{21}$

4. $\frac{16\sqrt{7}}{2} = \frac{8\sqrt{7}}{1} = 8\sqrt{7}$

5. $8\sqrt{2}$

6. 8

7. $7\sqrt{2}\sqrt{2} = 7\sqrt{4} = 7(2) = 14$

8. $7\sqrt{2}$

9. $\left(3\sqrt{10}\right)\left(7\sqrt{10}\right) = (3)(7)\sqrt{10}\sqrt{10} =$
   $21\sqrt{100} = 21(10) = 210$

10. $5\sqrt{7} + 4\sqrt{3} = 5\sqrt{7} + 4\sqrt{3}$:
    cannot be simplified

11. $\frac{10\sqrt{18}}{2\sqrt{6}} = \frac{10\sqrt{3}}{2} = \frac{5\sqrt{3}}{1} = 5\sqrt{3}$

12. $11^2 + L^2 = 22^2$
    $121 + L^2 = 484$
    $L^2 = 363$
    $L = \sqrt{363} = \sqrt{121}\sqrt{3} = 11\sqrt{3}$ units

13. $A = \frac{1}{2}bh = \frac{1}{2}(11)\left(11\sqrt{3}\right) =$
    $60.5\sqrt{3}$ units^2

14. one triangle:
    $A = \frac{1}{2}bh = \frac{1}{2}\left(5\sqrt{11}\right)(14) =$
    $35\sqrt{11}$ in^2
    six triangles:
    $A = (6)\left(35\sqrt{11}\right) = 210\sqrt{11}$ in^2

15. $V = Bh = \pi r^2 h \approx (3.14)\left(6^2\right)(4) =$
    452.16 ft^3

16. $(62)(452.16) = 28,033.92$ lb

17. $28,033.92 \div 2,000 \approx 14.02$ tons

18. $45° - 45° - 90°$ triangle: $2Y\sqrt{2}$

19. using Pythagorean theorem:
    $$A^2 + (3A)^2 = H^2$$
    $$A^2 + (3A)(3A) = H^2$$
    $$A^2 + 9A^2 = H^2$$
    $$10A^2 = H^2$$
    $$\sqrt{10A^2} = H = \sqrt{A^2}\sqrt{10} = A\sqrt{10}$$

20. $$L^2 + B^2 = \left(B\sqrt{10}\right)^2$$
    $$L^2 + B^2 = \left(B\sqrt{10}\right)\left(B\sqrt{10}\right)$$
    $$L^2 + B^2 = B^2\sqrt{100}$$
    $$L^2 + B^2 = 10B^2$$
    $$L^2 = 9B^2$$
    $$L = \sqrt{9B^2} = \sqrt{9}\sqrt{B^2} = 3B$$

Lesson Practice 22A

1. axiom; postulate
2. theorems
3. converses
4. congruent
5. bisector
6. congruent
7. congruent
8. 180°
9. parallel
10. 360°

11. If alternate interior angles are congruent, the two lines cut by a transversal are parallel.

12. If a quadrilateral has two pairs of parallel sides, it is a parallelogram.

13. In a triangle, if the leg squared plus the leg squared equals the hypotenuse squared, it is a right triangle.

14. If two perpendicular lines intersect, they form right angles.

15. A quadrilateral with only one pair of parallel sides is a trapezoid.

16. property of symmetry

17. reflexive property

18. transitive property

Lesson Practice 22B

1. assumes

2. proven

3. converse

4. isosceles

5. transversal; congruent

6. complementary

7. exterior

8. midpoint
   $\left(\text{or perpendicular bisector}\right)$

9. rectangle

10. square

11. The measures of supplementary angles add up to 180°.

12. A quadrilateral with two pairs of parallel sides is a parallelogram.

13. If alternate exterior angles are congruent, the lines cut by a transversal are parallel.

14. If two line segments are congruent, they are equal in length.

15. A polygon with two pairs of parallel sides and four congruent sides is a rhombus.

16. transitive property

17. property of symmetry

18. reflexive property

Systematic Review 22C

1. unproven

2. congruent

3. corresponding, alternate interior, and exterior angles are congruent.

4. $A + B > C$

5. interior angles

6. if $A = B$ then $B = A$

7. $5\sqrt{6}\,(2) = 10\sqrt{6}$

8. $5\sqrt{6}\sqrt{3} = 5\sqrt{18} = 5\sqrt{9}\sqrt{2} =$
   $5(3)\sqrt{2} = 15\sqrt{2}$

9. $\dfrac{13}{\sqrt{2}} = \dfrac{13\sqrt{2}}{\sqrt{2}\sqrt{2}} = \dfrac{13\sqrt{2}}{\sqrt{4}} = \dfrac{13\sqrt{2}}{2}$

10. $\dfrac{13\sqrt{2}}{2}$

11. $\dfrac{2\sqrt{3}}{\sqrt{5}} = \dfrac{2\sqrt{3}\sqrt{5}}{\sqrt{5}\sqrt{5}} = \dfrac{2\sqrt{15}}{\sqrt{25}} = \dfrac{2\sqrt{15}}{5}$

12. $L^2 + 13^2 = 17^2$
    $L^2 + 169 = 289$
    $L^2 = 120$
    $L = \sqrt{120} = \sqrt{4}\sqrt{30} = 2\sqrt{30}$

13. exterior angle $= 180° - 120° = 60°$
    $360° \div 60° = 6$ sides

14. circle: $C = 2\pi r \approx (2)(3.14)(10)$
    $= 62.8$ in
    sector is $\dfrac{150°}{360°} = \dfrac{5}{12}$ of circle:
    $\dfrac{5}{12}(62.8) \approx 26.17$ in

15. circle: $A = \pi r^2 \approx (3.14)\left(10^2\right)$
    $= 314$ in^2
    sector is $\dfrac{150°}{360°} = \dfrac{5}{12}$ of circle:
    $\dfrac{5}{12}(314) \approx 130.83$ in^2

16. one triangle:

$$A = \frac{1}{2}bh = \frac{1}{2}\left(3\sqrt{13}\right)(10) = 15\sqrt{13}$$

eight triangles:

$$A = (8)\left(15\sqrt{13}\right) = 120\sqrt{13} \approx 432.67 \text{ in}^2$$

17. $V = Bh = \pi r^2 h \approx (3.14)\left(7.5^2\right)(4)$

$$= 706.5 \text{ ft}^3$$

18. $(62)(706.5) = 43,803 \text{ lb}$

19. $43,803 \div 2,000 \approx 21.9 \text{ tons}$

20. cone: $V = \frac{1}{3}Bh = \frac{1}{3}\pi r^2 h \approx$

$$\frac{1}{3}(3.14)\left(14^2\right)(16.5) = 3,384.92 \text{ ft}^3$$

cylinder: $V = Bh = \pi r^2 h \approx$

$$(3.14)\left(14^2\right)(17) = 10,462.48 \text{ ft}^3$$

total: $3,384.92 + 10,462.48 =$

$$13,847.4 \text{ ft}^3$$

Systematic Review 22D

1. proven
2. congruent
3. isosceles
4. square
5. 360°
6. $A = A$
7. $5(2) = 10$
8. $5\sqrt{3}$
9. $9\sqrt{2}$
10. 9
11. $H^2 = 12^2 + 16^2$

$$H^2 = 144 + 256$$

$$H^2 = 400$$

$$H = \sqrt{400}$$

$$H = 20 \text{ units}$$

12. $A = \frac{1}{2}bh = \frac{1}{2}(12)(16) = 96 \text{ units}^2$
13. exterior angle $= 180° - 150° = 30°$

$360° \div 30° = 12$ sides

14. see drawing

15. one triangle:

$$A = \frac{1}{2}bh = \frac{1}{2}\left(6\sqrt{13}\right)(25) = 75\sqrt{13}$$

ten triangles:

$$A = 10\left(75\sqrt{13}\right) = 750\sqrt{13} \approx$$

$$2,704.16 \text{ in}^2$$

16. $V = Bh = \pi r^2 h \approx (3.14)\left(9^2\right)(4) =$

$$1,017.36 \text{ ft}^3$$

17. $(62)(1,017.36) = 63,076.32 \text{ lb}$

18. $63,076.32 \div 2,000 \approx 31.54 \text{ tons}$

19. "roof":

$$A = 2(12)(6.7) = 160.8 \text{ m}^2$$

triangles:

$$A = (2)\frac{1}{2}bh = (2)\frac{1}{2}(6)(6) = 36 \text{ m}^2$$

sides:

$$A = 2(6)(5) + 2(12)(5) = 180 \text{ m}^2$$

bottom:

$$A = (6)(12) = 72 \text{ m}^2$$

total:

$$A = 160.8 + 36 + 180 + 72 =$$

$$448.8 \text{ m}^2$$

20. prism: $V = Bh = \frac{1}{2}(6)(6)(12) =$

$$216 \text{ m}^3$$

rectangular solid:

$$V = (6)(12)(5) = 360 \text{ m}^3$$

total: $V = 216 + 360 = 576 \text{ m}^3$

Systematic Review 22E

1. rectangle
2. right angle
3. rhombus
4. supplementary
5. equal measures

6. If A = B and B = C then A = C

7. $\dfrac{16}{2} = 8$

8. $8\sqrt{3}$

9. $9\sqrt{2}$

10. 9

11. $\dfrac{\sqrt{6}}{4\sqrt{8}} = \dfrac{\sqrt{6}\sqrt{2}}{4\sqrt{8}\sqrt{2}} = \dfrac{\sqrt{12}}{4\sqrt{16}} = \dfrac{\sqrt{4}\sqrt{3}}{4(4)} =$
   $\dfrac{2\sqrt{3}}{16} = \dfrac{\sqrt{3}}{8}$

12. $48^2 + L^2 = 50^2$

   $2{,}304 + L^2 = 2{,}500$

   $L^2 = 196$

   $L = \sqrt{196} = 14$ units

13. circle:

   $C = 2\pi r \approx (2)(3.14)(6) = 37.68$ in

   sector is $\dfrac{360°}{360°}$ = whole circle

   $C = 1(37.68) = 37.68$ in

14. circle:

   $A = \pi r^2 \approx (3.14)(6^2) = 113.04$ in^2

   sector: $1(113.04) = 113.04$ in^2

15. one triangle:

   $A = \dfrac{1}{2}bh = \dfrac{1}{2}(8\sqrt{10})(22) = 88\sqrt{10}$ in^2

   five triangles:

   $A = (5)(88\sqrt{10}) = 440\sqrt{10} \approx$
   $1{,}391.40$ in^2

16. $V = Bh = \pi r^2 h \oplus (3.14)(12^2)(4) =$
   $1{,}808.64$ ft^3

17. $(62)(1{,}808.64) = 112{,}135.68$ lb
   $112{,}135.68 \div 2{,}000 \approx 56.07$ tons

18. $X^4 - 81 =$
   $(x^2 - 9)(x^2 + 9) =$
   $(x - 3)(x + 3)(x^2 + 9)$

19. $X^3 - 9X =$
   $(x)(x^2 - 9) =$
   $(x)(x - 3)(x + 3)$

20. $X^4 - 25X^2 =$
   $(x^2)(x^2 - 25) =$
   $(x^2)(x - 5)(x + 5)$

Lesson Practice 23A

1. $\angle D$
2. $\triangle DFE$
3. $\triangle DEF$
4. $\angle A$
5. $\overline{AC}$
6. $\overline{DF}$
7. $\angle XTY$
8. $\angle RST$
9. $\overline{YT}$
10. $\overline{TR}$
11. $\triangle YTX$
12. $\overline{XT}$
13. correspond
14. exterior
15. remote interior
16. $180° - 120° = 60°$
17. $180° - 89° = 91°$
18. $m\angle D = m\angle B + m\angle C$
   $60° + 91° = 151°$
   (remote interior angles)

Lesson Practice 23B

1. $\overline{BD}$
2. $\overline{BF}$
3. $\triangle BDF$
4. $\angle C$
5. $\angle AEC$
6. $\overline{AC}$
7. $\overline{BC}$
8. $\overline{CE}$
9. $\triangle BCA$
10. $\angle B$

11. $\angle E$

12. $\overline{DE}$

13. 180

14. exterior

15. B, C

16. $180° - 84° = 96°$

17. $180° - 132° = 48°$

18. $m\angle D = m\angle B + m\angle C$
    $96° + 48° = 144°$
    $\left(\text{remote interior angles}\right)$

Systematic Review 23C

1. $\overline{LM}$

2. $\overline{JG}$

3. $\triangle JHG$

4. $\angle MKL$

5. $\angle LMK$

6. $180° - 123° = 57°$

7. $180° - 110° = 70°$

8. $m\angle A = 180° - \left(m\angle B + m\angle C\right)$
   $m\angle A = 180° - \left(57° + 70°\right)$
   $m\angle A = 180° - 127° = 53°$

9. $30° - 60° - 90°$ triangle
   hypotenuse = short leg $\times$ 2
   $5 \times 2 = 10$

10. $30° - 60° - 90°$ triangle
    long leg = short leg $\times \sqrt{3}$
    $5 \times \sqrt{3} = 5\sqrt{3}$

11. parallel sides

12. regular

13. complementary angles

14. parallelogram

15. leg squared; leg squared

16. Exterior angles of a polygon
    add up to 360°. If each has
    a measure of 120°, there must
    be 360° ÷120, or 3 sides.
    Triangle

17. $V = \frac{4}{3}\pi r^3 \approx \frac{4}{3}(3.14)(4.3)^3 =$
    $\frac{4}{3}(3.14)(79.507) \approx 332.87$ cm^3

18. $SA = 4\pi r^2 \approx 4(3.14)(4.3)^2 =$
    $4(3.14)(18.49) \approx 232.23$ cm^2

19. area of circle with r = 4:
    $A = \pi r^2 \approx (3.14)(4)^2 =$
    $(3.14)(16) = 50.24$ ft^2
    45° is $\frac{1}{8}$ of 360°,
    so area of arc =
    $\frac{1}{8}$ of 50.24 ft^2 =
    $50.24 \div 8 = 6.28$ ft^2
    $V = Bh$ (area of base times height)
    $V = 6.28(5.2) \approx 32.66$ ft^3

20. 32.66 ft$^3\left(62 \text{ lb} / \text{ft}^3\right) =$
    2,024.92 lb
    $\left(\text{using rounded answer from } \#19\right)$

Systematic Review 23D

1. $\overline{ZA}$

2. $\overline{YX}$

3. $\angle XAZ$

4. $\triangle ZAX$

5. $\triangle AXZ$

6. $m\angle C = 180° - \left(52° + 59°\right)$
   $m\angle C = 180° - 111° = 69°$

7. $m\angle X = 180° - m\angle C$
   $m\angle X = 180° - 69° = 111°$

8. $m\angle Y = m\angle C + 52°$
   $\left(\text{remote interior angles}\right)$
   $m\angle Y = 69° + 52° = 121°$

9. $A^2 + 48^2 = 50^2$
   $A^2 + 2,304 = 2,500$
   $A^2 = 196$
   $A = 14$ units

10. $10^2 + 11^2 = B^2$

$100 + 121 = B^2$

$221 = B^2$

$B = \sqrt{221}$ units

11. quadrilateral or parallelogram

12. perpendicular

13. congruent

14. 180°

15. congruent

16. exterior angles add up to 360°; 360 ÷ 60 = 6 sides; hexagon

17. $A = \dfrac{major}{2} \times \dfrac{minor}{2} \ (\pi) \approx$

$\dfrac{10}{2} \times \dfrac{15}{2} \ (3.14) =$

$(5)(7.5)(3.14) = 117.75 \ m^2$

18. change 42 inches to feet:

$42 \div 12 = 3.5 \ ft$

$V = \dfrac{1}{3} Bh = \dfrac{1}{3} \pi r^2 h \approx$

$\dfrac{1}{3} (3.14)(3.5)^2 (15) =$

$\dfrac{1}{3} (3.14)(12.25)(15) \approx 192.33 \ ft^3$

19. $60° = \dfrac{1}{6}$ of 360° so area of base

will be $\dfrac{1}{6}$ that of the whole circle.

$A = \pi r^2 \approx (3.14)(7)^2 =$

$(3.14)(49) = 153.86 \ ft^2$

$153.86 \div 6 \approx 25.643 \ ft^2$

$V = Bh = 25.643 \times 2 \approx 51.29 \ ft^3$

20. $51.29 \ ft^3 \left(62 \ lb / ft^3\right) = 3,179.98 \ lb$

(using rounded answer from #19)

Systematic Review 23E

1. $\overline{VT}$

2. ∠QTV

3. ∠TVQ

4. △TSW

5. △SWT

6. $m\angle A = 180° - \left(62° + 67°\right) =$
$180° - \left(129°\right) = 51°$

7. $m\angle X = 180° - 67° = 113°$

8. $m\angle Y = 180° - 62° = 118°$

9. $A^2 + 34^2 = 36^2$

$A^2 + 1,156 = 1,296$

$A^2 = 140$

$A = \sqrt{140}$

$A = \sqrt{4}\sqrt{35} = 2\sqrt{35}$

10. $10^2 + 10^2 = B^2$

$100 + 100 = B^2$

$200 = B^2$

$B = \sqrt{200}$

$B = \sqrt{100}\sqrt{2} = 10\sqrt{2}$

11. vertex

12. $\dfrac{1}{2}$

13. alternate; congruent

14. 180°

15. >; The sum of the lengths of the two shorter sides of a triangle must be greater than the length of the long side.

16. Octogon has 8 sides. Each exterior angle must have a measure of 360° ÷ 8, or 45°, so each interior angle must have a measure of 180° – 45° = 135°.

17. .33 ft ×12 = 3.96 in

$V = Bh = \dfrac{1}{3} \pi r^2 h \approx$

$\dfrac{1}{3} (3.14)(3.96)^2 (7) \approx 114.89 \ in^3$

18. $X^2 + 6X + 8 = 0$

$(X + 4)(X + 2) = 0$

$X + 4 = 0 \qquad X + 2 = 0$

$X = -4 \qquad\quad X = -2$

19.
$$x^2 + 5x = -6$$
$$x^2 + 5x + 6 = 0$$
$$(x+3)(x+2) = 0$$

$$x + 3 = 0 \qquad x + 2 = 0$$
$$x = -3 \qquad\quad x = -2$$

20.
$$x^2 - 3x + 2 = 30$$
$$x^2 - 3x - 28 = 0$$
$$(x-7)(x+4) = 0$$

$$x + 4 = 0$$
$$x - 7 = 0 \qquad x = -4$$
$$x = 7$$

Lesson Practice 24A

Please note: In some cases, the student will be able to do the proof in a slightly different way from the way given in the answer key. Order of statements is not important except where one statement depends upon another, in which case the supporting statement will need to be made first.

1. given
2. definition of a square
3. reflexive property
4. SSS postulate
5. given
6. given
7. transversal; congruent
8. SAS postulate
9. $\overline{AB}$; $\overline{CB}$ order of 9 and 10 may be reversed
10. $\overline{AD}$; $\overline{CD}$
11. $\overline{BD}$; $\overline{BD}$
12. $\triangle ABD$; $\triangle CBD$
13. given
14. given
15. ABCD is a rhombus
16. $\overline{AD} \cong \overline{CB}$ (order of 16 and 17 may be reversed)
17. $\overline{AB} \cong \overline{CD}$
18. $\overline{BD} \cong \overline{BD}$
19. $\triangle ABD \cong \triangle CDB$

Lesson Practice 24B

1. given
2. given
3. $\overline{BD} \cong \overline{BD}$
4. definition of bisector
5. given
6. definition of a rectangle
7. alternate interior angles
8. reflexive property
9. SAS
10. given
11. definition of perpendicular
12. right angles are congruent
13. given
14. definition of midpoint
15. reflexive property
16. SAS postulate
17. given
18. given
19. $\overline{EC} \cong \overline{BC}$ (19 and 20 may be reversed)
20. $\overline{AC} \cong \overline{DC}$
21. $\triangle ABC \cong \triangle DEC$

Systematic Review 24C

1. definition of square
2. definition of square
3. reflexive property
4. $\triangle SQU \cong \triangle URS$
5. yes: The diagonal of the square is a transversal across the parallel sides of the square, making the angles named alternate interior angles.
6. 126°
7. 65°
8. 54°
9. 61°
10. GFE
11. $\overline{FE}$

12. The length of the hypotenuse of a 45° – 45° –90° triangle is equal to $\sqrt{2}$ times a leg of the triangle.
$$3\sqrt{2}\left(\sqrt{2}\right) = 3\left(\sqrt{4}\right) = 3(2) = 6$$

13. $3\sqrt{2}$ The legs of a 45° – 45° –90° triangle are congruent to one another.

14. The length of the hypotenuse of a 30° – 60° –90° triangle is equal to 2 times the short leg of the triangle.
$$2\sqrt{3}\,(2) = 4\sqrt{3}$$

15. The length of the long leg of a 30° – 60° –90° triangle is equal to $\sqrt{3}$ times the short leg of the triangle.
$$2\sqrt{3}\left(\sqrt{3}\right) = 2\left(\sqrt{9}\right) = 2(3) = 6$$

16. trapezoid

17. 180°

18. 2

19. Volume of cylinder :
$$V = Bh = \pi r^2 h \approx (3.14)(4)^2(8) = 401.92 \text{ ft}^3$$
Section is $\dfrac{90°}{360°} = \dfrac{1}{4}$ of the cylinder
$$\dfrac{1}{4} \times 401.92 = 100.48 \text{ ft}^3$$

20. $100.48 \text{ ft}^3 \times \dfrac{62 \text{ lb}}{\text{ft}^3} = 6,229.76 \text{ lb}$

8. $\angle DGE$

9. $\overline{GE}$

10. divide hypotenuse by $\sqrt{2}$:
$$\dfrac{4}{\sqrt{2}} = \dfrac{4\sqrt{2}}{\sqrt{2}\sqrt{2}} = \dfrac{4\sqrt{2}}{\sqrt{4}} = \dfrac{4\sqrt{2}}{2} = 2\sqrt{2}$$

11. same as A, so $2\sqrt{2}$

12. $A = \dfrac{1}{2}bh = \dfrac{1}{2}(4)(9) = 18 \text{ in}^2$

13. First, find hypotenuse:
$$9^2 + 4^2 = H^2$$
$$81 + 16 = H^2$$
$$97 = H^2$$
$$H = \sqrt{97} \oplus 9.8$$
$P = $ sum of sides $=$
$4 + 9 + 9.8 = 22.8 \text{ in}$

14. $\text{leg}^2 + \text{leg}^2 = \text{hypotenuse}^2$
or $A^2 + B^2 = C^2$

15. postulates

16. congruent

17. isosceles

18. square

19. Volume of cylinder:
$$V = Bh = \pi r^2 h \approx (3.14)(10)^2(12) = 3,768 \text{ ft}^3$$
Section is $\dfrac{120°}{360°} = \dfrac{1}{3}$ of the cylinder
$$\dfrac{1}{3} \times 3,768 \text{ ft}^3 = 1,256 \text{ ft}^3$$

20. $1,256 \text{ ft}^3 \times \dfrac{62 \text{ lb}}{\text{ft}^3} = 77,872 \text{ lb}$

Systematic Review 24D

1. definition of isosceles triangle
2. definition of bisector
3. reflexive property
4. $\triangle CDM \cong \triangle CFM$
5. yes; corresponding angles are congruent
6. 69°
7. 111°

Systematic Review 24E

1. $\overline{RH} \cong \overline{SB}$ (1 and 2 may be reversed.)
2. $\overline{RB} \cong \overline{SH}$
3. $\overline{BH} \cong \overline{BH}$
4. SSS postulate
5. yes; corresponding angles are congruent

6. 51°

7. 113°

8. ∠TWV

9. $\overline{VT}$

10. $10^2 + LL^2 = 20^2$

    $100 + LL^2 = 400$

    $LL^2 = 300$

    $LL = \sqrt{300}$

    $LL = \sqrt{100}\sqrt{3} = 10\sqrt{3}$

11. This triangle's sides follow the pattern of 30° – 60° –90° triangles, so ∠a = 60°

12. This triangle's sides follow the pattern of 30° – 60° –90° triangles, so ∠b = 30°

13. $A = \frac{1}{2}bh = \frac{1}{2}(10)(10\sqrt{3}) = 50\sqrt{3}$

    ≈ 86.60 square units

14. $P = 20 + 10 + 10\sqrt{3} = 30 + 10\sqrt{3}$

    ≈ 47.32 units

15. Exterior angles must add to 360°. Since there 10, each exterior angle must have a measure of $\frac{360°}{10}$ or 36°. Each interior angle, therefore will have a measure of 180° – 36° = 144°.

16. Arc BAC has a measure of 264°, so arc BC has a measure of 360° – 264° = 96°. The measure of the inscribed angle is half the measure of the arc,

    so m∠BAC = $\frac{1}{2}$ x96° = 48°

17. $\frac{3}{\sqrt{5}} + \frac{7}{\sqrt{2}} = \frac{3\sqrt{5}}{\sqrt{5}\sqrt{5}} + \frac{7\sqrt{2}}{\sqrt{2}\sqrt{2}} =$

    $\frac{3\sqrt{5}}{\sqrt{25}} + \frac{7\sqrt{2}}{\sqrt{4}} = \frac{3\sqrt{5}}{5} + \frac{7\sqrt{2}}{2} =$

    $\frac{(2)3\sqrt{5}}{(2)5} + \frac{(5)7\sqrt{2}}{(5)2} = \frac{6\sqrt{5} + 35\sqrt{2}}{10}$

18. $1,450,000 = 1.45 \times 10^6$

19. $.0076 = 7.6 \times 10^{-3}$

20. $640,000,000,000 = 6.4 \times 10^{11}$

Lesson Practice 25A

1. given

2. ∠BAD ≅ ∠BCD

3. given

4. $\overline{DB}$ bisects ∠ABC

5. definition of bisector

6. △ABD ≅ △CBD

7. ASA

8. yes: CPCTRC

9. ∠ABD ≅ ∠ACD

10. given

11. $\overline{AD}$ bisects ∠BAC

12. definition of bisector

13. $\overline{AD} ≅ \overline{AD}$

14. △ADB ≅ △ADC

15. AAS

16. yes: CPCTRC

17. $\overline{BD}$ bisects ∠ABC and ∠ADC

18. ∠ABD ≅ ∠CBD

    (18 and 19 may be reversed)

19. ∠ADB ≅ ∠CDB

20. $\overline{BD} ≅ \overline{BD}$

21. △BAD ≅ △BCD

22. ∠BAD ≅ ∠BCD

23. E is the midpoint of $\overline{FG}$

24. $\overline{EF} ≅ \overline{EG}$

25. definition of midpoint

26. $\overline{DG} \parallel \overline{HF}$

27. ∠DGE ≅ ∠HFE

28. ∠DEG ≅ ∠HEF

29. △DEG ≅ △HEF

30. ∠GDE ≅ ∠FHE

Lesson Practice 25B

1. given

2. $\overline{AB} ≅ \overline{CD}$

3. $\overline{AB} \parallel \overline{CD}$
4. $\triangle ABC \cong \triangle ADC$
5. $\overline{BD}$ bisects $\angle ABC$
6. given
7. definition of bisector
8. $\angle BAD \cong \angle BCD$
9. $\overline{BD} \cong \overline{BD}$
10. $\triangle ABD \cong \triangle CBD$
11. yes; CPCTRC
12. given
13. alternate interior angles
14. given
15. $\overline{HG} \cong \overline{FG}$
16. definition of bisector
17. $\angle HGM \cong \angle FGE$
18. $\triangle HGM \cong \triangle FGE$
19. yes: CPCTRC
20. given
21. alternate interior angles
22. $\angle BAF$ and $\angle EDC$ are right angles
23. $\angle BAF \cong \angle EDC$
24. $\overline{CE} \cong \overline{FB}$
25. $\triangle CDE \cong \triangle FAB$
26. AAS
27. yes: CPCTRC

Systematic Review 25C

1. $\angle QPT \cong \angle QSU$
   (1 and 2 may be reversed)
2. Q is the midpoint of $\overline{UT}$
3. $\overline{UQ} \cong \overline{TQ}$
   (3 and 5 may be reversed)
4. definition of midpoint
   (4 and 6 may be reversed if
   3 and 5 were reversed)
5. $\angle UQS \cong \angle TQP$
6. vertical angles
7. $\triangle QSU \cong \triangle QPT$
8. AAS
9. yes: CPCTRC
10. $180° - 123° = 57°$

11. $180° - 111° = 69°$
12. $180° - (57° + 69°) =$
    $180° - 126° = 54°$
13. QRS
14. $\overline{RS}$
15. If the square is viewed as two congruent triangles, one pointing up, and one pointing down, we can find the area of these **two** triangles, and add them to find the area of the square. Since the diagonals are 12 inches long, and bisect each other, we know that the height of each triangle is 6 inches, so the area of one triangle would be:

    $\frac{1}{2}bh = \frac{1}{2}(12)(6) = 36 \text{ in}^2$

    Area of both triangles combined:

    $36 \text{ in}^2 + 36 \text{ in}^2 = 72 \text{ in}^2$

    In problems 16 – 20, your answer may be slightly different from the one given, depending upon whether the intermediate steps were rounded or not. Answers that are close may be considered correct.

16. Looking at one of the four small triangles, each of the legs has a length of 6 in, so the hypotenuse of these triangles is:

    $6^2 + 6^2 = H^2$

    $36 + 36 = H^2$

    $72 = H^2$

    $H = \sqrt{72} \text{ or } 6\sqrt{2} \approx 8.49 \text{ in}$

    $P = 4(8.49) = 33.96 \text{ in}$

17. area of circle:

    $\pi r^2 \approx 3.14(8)^2 = 200.96 \text{ in}^2$

    $\text{sector} = \frac{36°}{360°} = \frac{1}{10} \text{ of the circle}$

    $\frac{1}{10}(200.96 \text{ in}^2) = 20.096 \text{ in}^2$

18. circumference of circle:

$$c = 2\pi r \approx 2(3.14)(8) = 50.24 \text{ in}$$

$$\text{sector} = \frac{36°}{360°} = \frac{1}{10} \text{ of the circle}$$

$$\frac{1}{10}(50.24) = 5.024 \text{ in}$$

19. 2 m = 200 cm

$$V = (200)(340)(560) =$$

$$38{,}080{,}000 \text{ cm}^3$$

or

340 cm = 3.4 m and 560 cm = 5.6 m

$$V = (2)(3.4)(5.6) = 38.08 \text{ m}^3$$

20. SA =

$$2(2)(3.4) + 2(2)(5.6) + 2(3.4)(5.6) =$$

$$13.6 + 22.4 + 38.08 = 74.08 \text{ m}^2$$

or

SA =

$$2(200)(340) + 2(200)(560) + 2(340)(560) =$$

$$136{,}000 + 224{,}000 + 380{,}800 =$$

$$740{,}800 \text{ cm}^2$$

Systematic Review 25D

Keep in mind the fact that the order of some steps in these proofs may be interchangeable.

1. $\overline{KZ}$ bisects $\angle JKX$
2. $\overline{KZ}$ bisects $\angle XZJ$
3. $\angle JKZ \cong \angle XKZ$
4. definition of bisector
5. $\angle XZK \cong \angle JZK$
6. definition of bisector
7. $\overline{KZ} \cong \overline{KZ}$
8. reflexive property
9. $\triangle JKZ \cong \triangle XKZ$
10. ASA
11. $\overline{KJ} \cong \overline{KX}$
12. $180° - 83° = 97°$
13. $180° - (28° + 97°) =$
    $180° - 125° = 55°$

14. $$S^2 + 6^2 = 15^2$$
    $$S^2 + 36 = 225$$
    $$S^2 = 189$$
    $$S = \sqrt{189} = \sqrt{9}\sqrt{21} = 3\sqrt{21} \approx 13.75$$

15. Base of top triangle = 6 cm

    Height of top triangle = $\frac{1}{2}(8) =$

        4 cm

    Area of top triangle = $\frac{1}{2}(6)(4) =$

        12 cm^2

    Top and bottom triangles are congruent, so:

    $$A = 12 + 12 = 24 \text{ cm}^2$$

16. The four small triangles each have sides measuring 4 cm and 3 cm, so the hypotenuse of each is:

    $$H^2 = 4^2 + 3^2$$
    $$H^2 = 16 + 9$$
    $$H^2 = 25$$
    $$H = 5$$
    Perimeter = $4(5) = 20$ cm

17. Volume of cylinder:

    $$V = Bh = \pi r^2 h \approx$$

    $$3.14(4)^2(3.3) \approx 165.79 \text{ cm}^3$$

    $$\text{section} = \frac{10°}{360°} = \frac{1}{36} \text{ of cylinder}$$

    $$\frac{1}{36}(165.79 \text{ cm}^3) \approx 4.61 \text{ cm}^3$$

18. Circumference of circle:

    $$C = 2\pi r \approx 2(3.14)(4) = 25.12 \text{ cm}$$

    $$\text{arc} = \frac{10°}{360°} = \frac{1}{36} \text{ of circle}$$

    $$\frac{1}{36}(25.12) \approx .70 \text{ cm}$$

19. convert all measurements to yards:

$$\frac{6 \text{ ft}}{1} \times \frac{1 \text{ yd}}{3 \text{ ft}} = \frac{6 \text{ yd}}{3} = 2 \text{ yd}$$

$$\frac{3\sqrt{3} \text{ ft}}{1} \times \frac{1 \text{ yd}}{3 \text{ ft}} = \frac{3\sqrt{3} \text{ yd}}{3} =$$

$$\frac{\sqrt{3} \text{ yd}}{1} = \sqrt{3} \text{ yd}$$

volume = area of base times height. Triangular end is base, so:

$$V = Bh = \frac{1}{2}(2)(\sqrt{3})(5)$$

$$= 5\sqrt{3} \approx 8.66 \text{ yd}^3$$

20. SA = area of three rectangular sides, plus area of two triangular ends. Note that the ends are equilateral triangles. This can be verified by using the pythagorean theorem and the information given.

SA =

$$3(5 \text{ yd})(2 \text{ yd}) + 2(\frac{1}{2})(2 \text{ yd})(\sqrt{3} \text{ yd}) =$$

$$30 \text{ yd}^2 + 2\sqrt{3} \text{ yd}^2 =$$

$$30 + 2\sqrt{3} \text{ yd}^2 \approx 33.46 \text{ yd}^2$$

Systematic Review 25E

1. $\angle LPG \cong \angle RGP$
2. alternate interior angles
3. $\angle RPG \cong \angle LGP$
4. alternate interior angles
5. $\overline{PG} \cong \overline{PG}$
6. reflexive property
7. $\triangle PLG \cong \triangle GRP$
8. ASA
9. yes: CPCTRC
10. Exterior angles of a polygon add up to 360°:
   $360° - (115° + 119°) =$
   $360° - 234° = 126°$

11. $180° - 115° = 65°$
12. $180° - 126° = 54°$
    (126° from answer #10)
13. $180° - 119° = 61°$
14. Top triangle:
    $A = \frac{1}{2}bh = \frac{1}{2}(10)(6) = 30 \text{ in}^2$
    Bottom triangle:
    $A = \frac{1}{2}bh = \frac{1}{2}(10)(10) = 50 \text{ in}^2$
    Total area:
    $30 \text{ in}^2 + 50 \text{ in}^2 = 80 \text{ in}^2$
15. One of two small triangles:
    $h^2 = 6^2 + 5^2$
    $h^2 = 36 + 25$
    $h^2 = 61$
    $h = \sqrt{61} \approx 7.81 \text{ in}$
    One of two large triangles:
    $h^2 = 10^2 + 5^2$
    $h^2 = 100 + 25$
    $h^2 = 125$
    $h = \sqrt{125} \approx 11.18 \text{ in}$
    $P = 2(7.81) + 2(11.18) =$
    $15.62 + 22.36 = 37.98 \text{ in}$
16. circle :
    $A = \pi r^2 \approx 3.14(7)^2 = 153.86 \text{ mm}^2$
    $\text{sector} = \frac{240°}{360°} = \frac{2}{3} \text{ of circle}$
    $\frac{2}{3}(153.86 \text{ mm}^2) \approx 102.57 \text{ mm}^2$
17. circle :
    $C = 2\pi r \approx 2(3.14)(7) = 43.96 \text{ mm}$
    $\text{sector} = \frac{240°}{360°} = \frac{2}{3} \text{ of circle}$
    $\frac{2}{3}(43.96 \text{ mm}) \approx 29.31 \text{ mm}$
18. $5,000 \times 8,000,000 =$
    $(5 \times 10^3)(8 \times 10^6) =$
    $(5 \times 8)(10^3 \times 10^6) =$
    $40 \times 10^9 = 4 \times 10^{10}$

19. $18,000 \times .007 =$

$(1.8 \times 7.0)(10^4 \times 10^{-3}) =$

$(12.6)(10^1) = 12.6 \times 10^1 =$

1.26×10^2

or: 1×10^2

if the student took significant digits into account.

20. $1,400,000 \div 290 =$

$(1.4 \times 10^6) \div (2.9 \times 10^2) =$

$(1.4 \div 2.9)(10^6 \div 10^2) =$

$.48 \times 10^4 = 4.8 \times 10^3$

Lesson Practice 26A

1. $\overline{AC} \perp \overline{BD}$
2. definition of perpendicular
3. definition of perpendicular
4. $\triangle ABD$ is isosceles
5. $\overline{AB} \cong \overline{AD}$
6. $\overline{AC} \cong \overline{AC}$
7. $\triangle ABC \cong \triangle ADC$
8. CPCTRC
9. C is the midpoint of $\overline{BD}$
10. definition of midpoint
11. $\triangle HJM$ is equilateral
12. angles in an equilateral triangle are congruent
13. $\overline{HK} \perp \overline{JM}$
14. definition of perpendicular
15. definition of perpendicular
16. $\overline{HK} \cong \overline{HK}$
17. $\triangle HJK \cong \triangle HMK$
18. ABCD is a rhombus
19. $\overline{AB} \cong \overline{CD}$
20. $\overline{AC} \perp \overline{DB}$
21. given
22. $\angle AOB$ is a right angle
23. $\angle COD$ is a right angle
24. O is the midpoint of $\overline{BD}$

25. $\overline{OD} \cong \overline{OB}$
26. definition of midpoint
27. $\triangle ABO \cong \triangle CDO$
28. HL

Lesson Practice 26B

1. $\overline{HE} \perp \overline{FC}$
2. $\overline{BG} \perp \overline{FC}$
3. $\angle BGC$ is a right angle
4. $\angle EHF$ is a right angle
5. $\overline{HE} \cong \overline{GB}$
6. given
7. $\overline{FE} \cong \overline{CB}$
8. given
9. $\triangle FHE \cong \triangle CGB$
10. HL
11. $\overline{AC}$ is tangent to circle at B
12. $\angle DBA$ is a right angle
13. property of tangent (see lesson 12)
14. $\angle DBC$ is a right angle
15. property of tangent
16. B is the midpoint of $\overline{AC}$
17. $\overline{AB} \cong \overline{CB}$
18. definition of midpoint
19. $\overline{DB} \cong \overline{DB}$
20. reflexive property
21. $\triangle ABD \cong \triangle CBD$
22. LL
23. $\overline{AG}$ is tangent to circle at G
24. $\overline{DE}$ is tangent to circle at E
25. $\overline{AG} \perp \overline{GF}$
26. property of tangent
27. $\overline{DE} \perp \overline{EF}$
28. property of tangent
29. $\angle AGF$ is a right angle
30. $\angle DEF$ is a right angle
31. $\angle GAF \cong \angle EDF$
32. given
33. $\overline{GF} \cong \overline{EF}$

34. If two line segments have equal lengths, they are congruent. Radii of a circle have equal lengths.
35. $\triangle AGF \cong \triangle DEF$
36. LA
37. $\overline{AF} \cong \overline{DF}$
38. CPCTRC

Systematic Review 26C

1. $\overline{RL} \cong \overline{GN}$ or $\overline{RN} \cong \overline{GL}$
2. opposite sides of a rectangle are congruent (APT)
3. $m\angle NRL \cong m\angle LGN = 90°$
4. opposite sides of a rectangle are congruent (APT)
5. $\overline{LN} \cong \overline{LN}$
6. reflexive property
7. $\triangle RNL \cong \triangle GLN$
8. HL

9. $$L^2 + \left(6\sqrt{2}\right)^2 = 12^2$$
$$L^2 + (6)(6)\left(\sqrt{2}\right)\left(\sqrt{2}\right) = 144$$
$$L^2 + (36)(2) = 144$$
$$L^2 + 72 = 144$$
$$L^2 = 72$$
$$L = \sqrt{72}$$
$$L = \sqrt{36}\sqrt{2}$$
$$L = 6\sqrt{2}$$
or ≈ 8.49 units

10. $$6^2 + 9^2 = H^2$$
$$36 + 81 = H^2$$
$$117 = H^2$$
$$H = \sqrt{117} \text{ or} \approx 10.82 \text{ units}$$

11. $180°$
12. 2
13. $\sqrt{2}$

14. $$\frac{2}{\sqrt{5}} \times \frac{5}{\sqrt{2}} = \frac{10}{\sqrt{10}} =$$
$$\frac{10}{\sqrt{10}} \times \frac{\sqrt{10}}{\sqrt{10}} = \frac{10\sqrt{10}}{\sqrt{100}} =$$
$$\frac{10\sqrt{10}}{10} = \sqrt{10}$$

15. $C = 2\pi r \approx 2(3.14)(2.3) \approx 14.44$ cm
16. $A = \pi r^2 \approx (3.14)(2.3)^2 \approx 16.61$ cm^2
17. Exterior angles add up to $360°$, so the measure of each exterior angle is $\frac{360°}{15} = 24°$.
interior angles $= 180° - 24° = 156°$
18. point
19. $123°$
20. Alternate exterior angles are congruent.

Systematic Review 26D

1. $\overline{PT} \cong \overline{PS}$
2. definition of midpoint
3. $\overline{PQ} \cong \overline{PQ}$
4. reflexive property
5. $\triangle PTQ \cong \triangle PSQ$
6. LL
7. $\overline{SQ} \cong \overline{TQ}$
8. CPCTRC
9. The perpendicular bisector will divide the triangle into two smaller triangles, each of which will have a base of 4 inches, because of the definition of bisector. Using one of these small triangles, we have a leg of 4 and a hypotenuse of 8. Using L for the unknown leg :
$$L^2 + 4^2 = 8^2$$
$$L^2 + 16 = 64$$
$$L^2 = 48$$
$$L = \sqrt{48}$$
$$L = \sqrt{16}\sqrt{3}$$
$$L = 4\sqrt{3} \text{ in}$$

10. $A = \frac{1}{2}bh = \frac{1}{2}(8)(4\sqrt{3}) =$
$(4)(4\sqrt{3}) = 16\sqrt{3}$ in^2

11. Each of the smaller triangles are $45^\circ - 45^\circ - 90^\circ$ triangles, so they both have a pair of legs with equal measures. Therefore, the base of the larger triangle has a measure of $7 + 7 = 14$ inches. The hypotenuse of each of the smaller triangles is $\sqrt{2}$ times the leg, or $7\sqrt{2}$.
$P = S + S + S = 14 + 7\sqrt{2} + 7\sqrt{2} =$
$14 + 14\sqrt{2} \approx 33.80$ in

12. $A = \frac{1}{2}bh = \frac{1}{2}(14)(7) = 49$ in^2

13. 8

14. $\frac{\sqrt{5}}{3\sqrt{2}} \times \frac{\sqrt{3}}{4\sqrt{2}} = \frac{\sqrt{15}}{12\sqrt{4}} =$
$\frac{\sqrt{15}}{12(2)} = \frac{\sqrt{15}}{24}$

15. $V = \frac{4}{3}\pi r^3 \approx \frac{4}{3}(3.14)(4)^3 \approx$
267.95 in^3

16. $A = 4\pi r^2 \approx 4(3.14)(4)^2$
$= 200.96$ in^2

17. Exterior angles add up to 360°, so the measure of each exterior angle $= \frac{360^\circ}{18} = 20^\circ$.
Interior angles $= 180^\circ - 20^\circ = 160^\circ$

18. line or line segment or ray

19. $m\angle A = 180^\circ - 72^\circ = 108^\circ$

20. They are supplementary angles.

5. $\triangle RSQ \cong \triangle TSQ$

6. LA

7. $\overline{RQ} \cong \overline{TQ}$

8. CPCTRC

9. convert 30 cm to m:
$\frac{30}{100} = .3$
$V = Bh = \pi r^2 h \approx 3.14(.3)^2(4) \approx$
1.13 m^3

10. $SA = 2\pi r^2 + 2\pi rh \approx$
$2(3.14)(.3)^2 + 2(3.14)(.3)(4) \approx$
$.57 + 7.54 = 8.11$ m^2
Your answer may differ slightly from this, depending upon how and when you rounded.

11. side-side-side

12. side-angle-side

13. angle-side-angle

14. angle-angle-side

15. hypotenuse-leg

16. leg-leg

17. hypotenuse-angle

18. leg-angle

19. $A = \pi r^2 \cong 3.14(2X)^2 = 3.14(2X)(2X) =$
$3.14(4X^2) = 12.56X^2$ units^2

20. $(2X)^2 + (3X)^2 = H^2$
$4X^2 + 9X^2 = H^2$
$13X^2 = H^2$
$\sqrt{13X^2} = H$
$H = \sqrt{X^2}\sqrt{13} = X\sqrt{13}$ units

Systematic Review 26E

1. $\angle RQS \cong \angle TQS$

2. definition of bisector

3. $\overline{QS} \cong \overline{QS}$

4. reflexive property

Lesson Practice 27A

1. $\overline{AB} \perp \overline{BD}$

2. $\overline{EC} \perp \overline{BD}$

3. $\angle ABD \cong \angle ECD$

4. definition of perpendicular

5. $\angle EDC \cong \angle ADB$

6. reflexive property

7. △ABD ~ △ECD

8. AA

9. $\overline{AB} \parallel \overline{CD}$

10. ∠ABE ≅ ∠CDE

11. alternate interior angles

12. ∠AEB ≅ ∠CED

13. vertical angles

14. △ABE ~ △CDE

15. AA

For problems using proportions, there is often more than one possible way to set up the proportion. The student can use any method that results in a correct answer.

16. $\dfrac{5}{10} = \dfrac{8}{X}$

$5X = 80$

$X = 16$

17. $\dfrac{10}{15} = \dfrac{X}{24}$

$\dfrac{240}{15} = X$

$X = 16$

18. $\dfrac{15}{25} = \dfrac{6}{X}$

$15X = 150$

$X = 10$

19. $\dfrac{8}{10} = \dfrac{X}{13}$

$\dfrac{104}{10} = X$

$X = 10.4$

20. $\dfrac{5}{X} = \dfrac{6}{10}$

$50 = 6X$

$X = \dfrac{50}{6} = 8\dfrac{1}{3}$

Lesson Practice 27B

1. ∠DAB ≅ ∠EBC

2. ∠ACD ≅ ∠BCE

3. reflexive property

4. △DAC ~ △EBC

5. AA

6. ∠ABC ≅ ∠ADB

7. given

8. ∠DAB ≅ ∠BAC

9. reflexive property

10. △ABC ~ △ADB

11. AA

12. $\dfrac{X}{10} = \dfrac{5}{8}$

$8X = 50$

$X = \dfrac{50}{8} = 6\dfrac{1}{4}$ or 6.25

13. $\dfrac{X}{4} = \dfrac{9}{6}$

$6X = 36$

$X = 6$

14. $\dfrac{X}{15} = \dfrac{33}{25}$

$25X = 495$

$X = \dfrac{495}{25} = 19\dfrac{4}{5}$ or 19.8

15. $\dfrac{X}{13} = \dfrac{6}{15}$

$15X = 78$

$X = \dfrac{78}{15} = 5\dfrac{1}{5}$ or 5.2

16. $\dfrac{X}{6} = \dfrac{15}{10}$

$10X = 90$

$X = 9$

17. $\dfrac{(X+5)}{5} = \dfrac{16}{10}$

$10(X+5) = 80$

$10X + 50 = 80$

$10X = 30$

$X = 3$

Systematic Review 27C

1. ∠FGH ≅ ∠FJK

2. definition of perpendicular

3. ∠GFH ≅ ∠JFK

4. reflexive property

5. △FGH ~ △FJK
6. AA
7. ∠RCT and ∠ETC are right angles
8. definition of rectangle
9. $\overline{RC} \cong \overline{ET}$
10. definition of rectangle
11. $\overline{CT} \cong \overline{CT}$
12. reflexive property
13. △ECT ≅ △RTC
14. LL
15. yes: CPCTRC
16. $\dfrac{X}{10} = \dfrac{15}{25}$

    $25X = 150$

    $X = 6$
17. exterior angles add up to 360°

    $\dfrac{360°}{36°} = 10$ sides: decagon
18. A decagon can be divided into 10 congruent triangles. The area of one of these triangles would be:

    $A = \dfrac{1}{2}bh = \dfrac{1}{2}(5)(3\sqrt{2}) = 7.5\sqrt{2}$ cm^2

    $7.5\sqrt{2}$ (10 triangles) $= 75\sqrt{2}$ cm^2

    ≈ 106.07 cm^2
19. $180° - 105° = 75°$: supplementary angles
20. $m\angle\alpha = 180° - (43° + 75°) = 180° - 118° = 62°$

Systematic Review 27D

1. ∠AEB ≅ ∠DCB
2. alternate interior angles
3. ∠ABE ≅ ∠DBC
4. vertical angles
5. △AEB ~ △DCB
6. AA
7. $\overline{AB} \cong \overline{CB}$
8. given
9. $\overline{BD} \cong \overline{BD}$

10. reflexive property
11. △ABD ≅ △CBD
12. HL
13. $\overline{AD} \cong \overline{CD}$
14. CPCTRC
15. $\dfrac{X}{5} = \dfrac{30}{10}$

    $10X = 150$

    $X = 15$
16. exterior angles add up to 360°:

    $\dfrac{360°}{45°} = 8$ sides: octagon
17. Area of one triangle:

    $A = \dfrac{1}{2}bh = \dfrac{1}{2}(10)(12) = 60$ m^2

    Area of octagon:

    60 m^2 (8 triangles) $= 480$ m^2
18. $P = 8(10) = 80$ m
19. $180° - 99° = 81°$: supplementary angles
20. $m\angle A = 180° - (33° + 81°) = 180° - 114° = 66°$

Systematic Review 27E

1. ∠1 ≅ ∠2
2. given
3. ∠MLN ≅ ∠QPR
4. two angles with the same measure are congruent
5. △LMN ~ △PQR
6. AA
7. $\overline{CF} \cong \overline{CD}$
8. definition of isosceles triangle
9. ∠FCM ≅ ∠DCM
10. definition of bisector
11. $\overline{CM} \cong \overline{CM}$
12. reflexive property
13. △CDM ≅ △CFM
14. SAS

15. $\dfrac{(X+12)}{12} = \dfrac{20}{15}$

$15(X+12) = 240$

$15X + 180 = 240$

$15X = 60$

$X = 4$

16. Area of circle:

$A = \pi r^2 \approx 3.14(6)^2 = 113.04 \text{ ft}^2$

Sector is $\dfrac{330^\circ}{360^\circ} = \dfrac{11}{12}$ of circle

$\dfrac{11}{12}\left(113.04 \text{ ft}^2\right) = 103.62 \text{ ft}^2$

17. Volume of cylinder:

$V = Bh = \pi r^2 h \approx$

$3.14(6)^2(4.5) = 508.68 \text{ ft}^3$

Sector is $\dfrac{330^\circ}{360^\circ} = \dfrac{11}{12}$ of cylinder

$\dfrac{11}{12}\left(508.68 \text{ ft}^3\right) = 466.29 \text{ ft}^3$

18. $2X^\circ + 3X^\circ = 180^\circ$

(supplementary angles)

$5X^\circ = 180^\circ$

$X = 36^\circ$

19. $4X^\circ + X^\circ = 90^\circ$

(complementary angles)

$5X^\circ = 90^\circ$

$X = 18^\circ$

20. $P = S + S + S = 2X + 6X + 4X$

$2X + 6X + 4X = 12$

$12X = 12$

$X = 1$

Lesson Practice 28A

Here are some ideas to help with transformational geometry. For reflection problems, try placing the edge of a hand mirror on the axis of reflection, facing the figure to be reflected. The reflection in the mirror will show the image that would be the result of geometric reflection.

For rotation problems, lay a piece of tracing paper over the graph, trace the original figure, and plot the point of rotation. Without moving the tracing paper, insert a pin through the point of rotation and into the graph on the student page. Now, rotate the tracing paper counterclockwise the indicated number of degrees, and the new position of the image on the tracing paper will indicate the result of geometric rotation.

1.

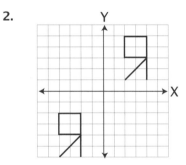

2.

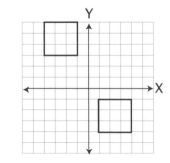

3.

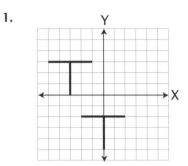

4.

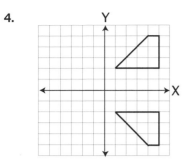

8.

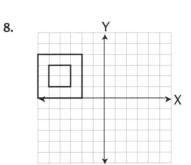

5.

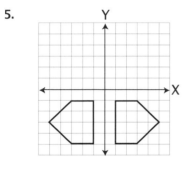

9.

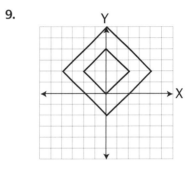

6.

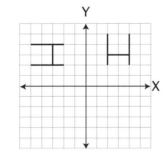

Lesson Practice 28B

1.

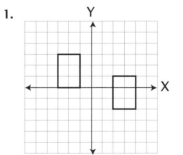

7.

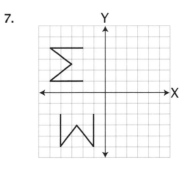

2.

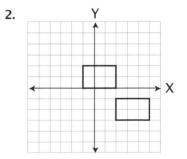

3.

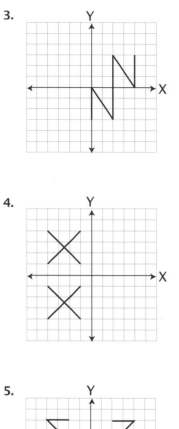

7.

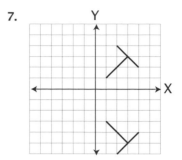

4.

8.

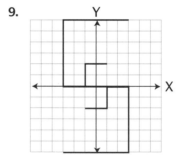

5.

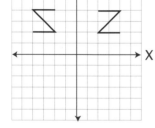

9.

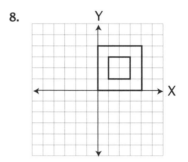

6.

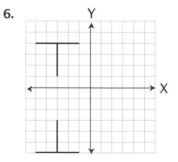

Systematic Review 28C

1- 4.

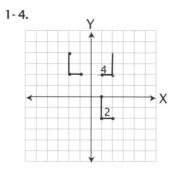

5-8.

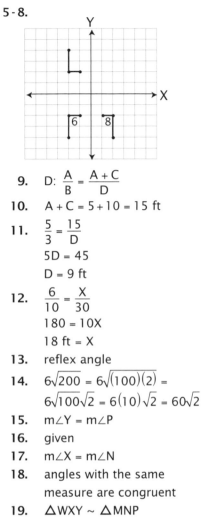

9. D: $\dfrac{A}{B} = \dfrac{A+C}{D}$

10. $A + C = 5 + 10 = 15$ ft

11. $\dfrac{5}{3} = \dfrac{15}{D}$

$5D = 45$

$D = 9$ ft

12. $\dfrac{6}{10} = \dfrac{X}{30}$

$180 = 10X$

18 ft $= X$

13. reflex angle

14. $6\sqrt{200} = 6\sqrt{(100)(2)} =$
$6\sqrt{100}\sqrt{2} = 6(10)\sqrt{2} = 60\sqrt{2}$

15. $m\angle Y = m\angle P$

16. given

17. $m\angle X = m\angle N$

18. angles with the same measure are congruent

19. $\triangle WXY \sim \triangle MNP$

20. AA

Systematic Review 28D

1-4.

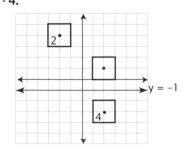

5.

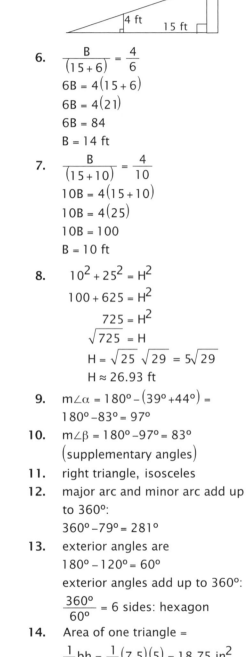

6. $\dfrac{B}{(15+6)} = \dfrac{4}{6}$

$6B = 4(15+6)$

$6B = 4(21)$

$6B = 84$

$B = 14$ ft

7. $\dfrac{B}{(15+10)} = \dfrac{4}{10}$

$10B = 4(15+10)$

$10B = 4(25)$

$10B = 100$

$B = 10$ ft

8. $10^2 + 25^2 = H^2$

$100 + 625 = H^2$

$725 = H^2$

$\sqrt{725} = H$

$H = \sqrt{25}\,\sqrt{29} = 5\sqrt{29}$

$H \approx 26.93$ ft

9. $m\angle\alpha = 180° - (39° + 44°) =$
$180° - 83° = 97°$

10. $m\angle\beta = 180° - 97° = 83°$
(supplementary angles)

11. right triangle, isosceles

12. major arc and minor arc add up to 360°:
$360° - 79° = 281°$

13. exterior angles are
$180° - 120° = 60°$
exterior angles add up to 360°:
$\dfrac{360°}{60°} = 6$ sides: hexagon

14. Area of one triangle $=$
$\dfrac{1}{2}bh = \dfrac{1}{2}(7.5)(5) = 18.75$ in^2
Six triangles:
$6(18.75$ in$^2) = 112.5$ in^2

15. $P = 6(7.5$ in$) = 45$ in

16. ▱RHSB is a rhombus

17. $\overline{BS} \cong \overline{RH}$: definition of a rhombus

18. $\overline{BR} \cong \overline{HS}$: definition of a rhombus

19. $\overline{BH} \cong \overline{BH}$: reflexive property

20. $\triangle HSB \cong \triangle BRH$: SSS

Systematic Review 28E
1- 4.

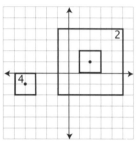

5.

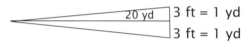

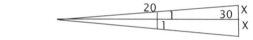

6. $\dfrac{X}{1} = \dfrac{50}{20}$

$20X = 50$

$X = \dfrac{50}{20} = 2\dfrac{1}{2}$ or 2.5 yd

(or 7.5 ft)

7. $50^2 + 2.5^2 = H^2$

$2{,}500 + 6.25 = H^2$

$2{,}506.25 = H^2$

$H = \sqrt{2{,}506.25} \approx 50.06$ yd (or 150.2 ft)

8. Exterior angle $= 180° - 108° = 72°$

$\dfrac{360°}{72°} = 5$: pentagon

9. Area of one triangle:

$A = \dfrac{1}{2}bh = \dfrac{1}{2}(11)(8) = 44 \text{ m}^2$

Area of five triangles:

$44 \text{ m}^2 (5) = 220 \text{ m}^2$

10. $P = 5(11 \text{ m}) = 55 \text{ m}$

11. $\overline{KZ}$ bisects $\angle JKX$ and $\angle JZX$

(Numbers 12-14 may be in any order.)

12. $\angle JKZ \cong \angle XKZ$:

definition of bisector

13. $\angle JZK \cong \angle XZK$:

definition of bisector

14. $\overline{KZ} \cong \overline{KZ}$: reflexive property

15. $\triangle JKZ \cong \triangle XKZ$: ASA

16. $P = 2(X+3) + 2(X+1)$

$2(X+3) + 2(X+1) = 36$

$2X + 6 + 2X + 2 = 36$

$4X + 8 = 36$

$4X = 28$

$X = 7$ ft

17. $A = bh = (X+3)(X)$

$(X+3)(X) \Rightarrow (7+3)(7) = (10)(7) = 70 \text{ ft}^2$

18. $P = 2(Y+4) + 2(Y-1)$

$2(Y+4) + 2(Y-1) = 98$

$2Y + 8 + 2Y - 2 = 98$

$4Y + 6 = 98$

$4Y = 92$

$Y = 23$ in

19. $A = bh = (Y+4)(Y-1)$

$(Y+4)(Y-1) \Rightarrow$

$((23)+4)((23)-1) =$

$(27)(22) = 594 \text{ in}^2$

20. $5X° + 4X° = 90°$:

complementary angles

$9X° = 90°$

$X = 10°$

Lesson Practice 29A

1. $\dfrac{5}{13}$

2. $\dfrac{12}{13}$

3. $\dfrac{5}{12}$

4. $\dfrac{12}{13}$

5. $\dfrac{5}{13}$

6. $\dfrac{12}{5}$

7. $\dfrac{7}{25}$

8. $\dfrac{24}{25}$

9. $\dfrac{7}{24}$

10. $\dfrac{24}{25}$

11. $\dfrac{7}{25}$

12. $\dfrac{24}{7}$

13. $\dfrac{6}{18} = \dfrac{1}{3}$

14. $\dfrac{17}{18}$

15. $\dfrac{6}{17}$

16. $\dfrac{17}{18}$

17. $\dfrac{6}{18} = \dfrac{1}{3}$

18. $\dfrac{17}{6}$

19. $\dfrac{8}{10} = \dfrac{4}{5}$

20. $\dfrac{6}{10} = \dfrac{3}{5}$

21. $\dfrac{8}{6} = \dfrac{4}{3}$

22. $\dfrac{6}{10} = \dfrac{3}{5}$

23. $\dfrac{8}{10} = \dfrac{4}{5}$

24. $\dfrac{6}{8} = \dfrac{3}{4}$

Lesson Practice 29B

1. $\dfrac{5}{5\sqrt{2}} = \dfrac{1}{\sqrt{2}} \times \dfrac{\sqrt{2}}{\sqrt{2}} = \dfrac{\sqrt{2}}{2}$

2. $\dfrac{5}{5\sqrt{2}} = \dfrac{1}{\sqrt{2}} \times \dfrac{\sqrt{2}}{\sqrt{2}} = \dfrac{\sqrt{2}}{2}$

3. $\dfrac{5}{5} = 1$

4. $\dfrac{5}{5\sqrt{2}} = \dfrac{1}{\sqrt{2}} \times \dfrac{\sqrt{2}}{\sqrt{2}} = \dfrac{\sqrt{2}}{2}$

5. $\dfrac{5}{5\sqrt{2}} = \dfrac{1}{\sqrt{2}} \times \dfrac{\sqrt{2}}{\sqrt{2}} = \dfrac{\sqrt{2}}{2}$

6. $\dfrac{5}{5} = 1$

7. $\dfrac{13}{26} = \dfrac{1}{2}$

8. $\dfrac{13\sqrt{3}}{26} = \dfrac{\sqrt{3}}{2}$

9. $\dfrac{13}{13\sqrt{3}} = \dfrac{1}{\sqrt{3}} \times \dfrac{\sqrt{3}}{\sqrt{3}} = \dfrac{\sqrt{3}}{3}$

10. $\dfrac{13\sqrt{3}}{26} = \dfrac{\sqrt{3}}{2}$

11. $\dfrac{13}{26} = \dfrac{1}{2}$

12. $\dfrac{13\sqrt{3}}{13} = \dfrac{\sqrt{3}}{1} = \sqrt{3}$

13. $\dfrac{7}{\sqrt{130}} \times \dfrac{\sqrt{130}}{\sqrt{130}} = \dfrac{7\sqrt{130}}{130}$

14. $\dfrac{9}{\sqrt{130}} \times \dfrac{\sqrt{130}}{\sqrt{130}} = \dfrac{9\sqrt{130}}{130}$

15. $\dfrac{7}{9}$

16. $\dfrac{9}{\sqrt{130}} \times \dfrac{\sqrt{130}}{\sqrt{130}} = \dfrac{9\sqrt{130}}{130}$

17. $\dfrac{7}{\sqrt{130}} \times \dfrac{\sqrt{130}}{\sqrt{130}} = \dfrac{7\sqrt{130}}{130}$

18. $\dfrac{9}{7}$

19. $\dfrac{\sqrt{203}}{18}$

20. $\dfrac{11}{18}$

21. $\dfrac{\sqrt{203}}{11}$

22. $\dfrac{11}{18}$

23. $\dfrac{\sqrt{203}}{18}$

24. $\dfrac{11}{\sqrt{203}} \times \dfrac{\sqrt{203}}{\sqrt{203}} = \dfrac{11\sqrt{203}}{203}$

Systematic Review 29C

1. $\dfrac{30}{50} = \dfrac{3}{5}$

2. $\dfrac{40}{50} = \dfrac{4}{5}$

3. $\dfrac{30}{40} = \dfrac{3}{4}$

4. $\dfrac{40}{50} = \dfrac{4}{5}$

5. $\dfrac{30}{50} = \dfrac{3}{5}$

6. $\dfrac{40}{30} = \dfrac{4}{3}$

7. $\dfrac{A}{C}$

8. $\dfrac{B}{C}$

9. $\dfrac{A}{B}$

10. $\dfrac{B}{C}$

11. $\dfrac{A}{C}$

12. $\dfrac{B}{A}$

13-16.

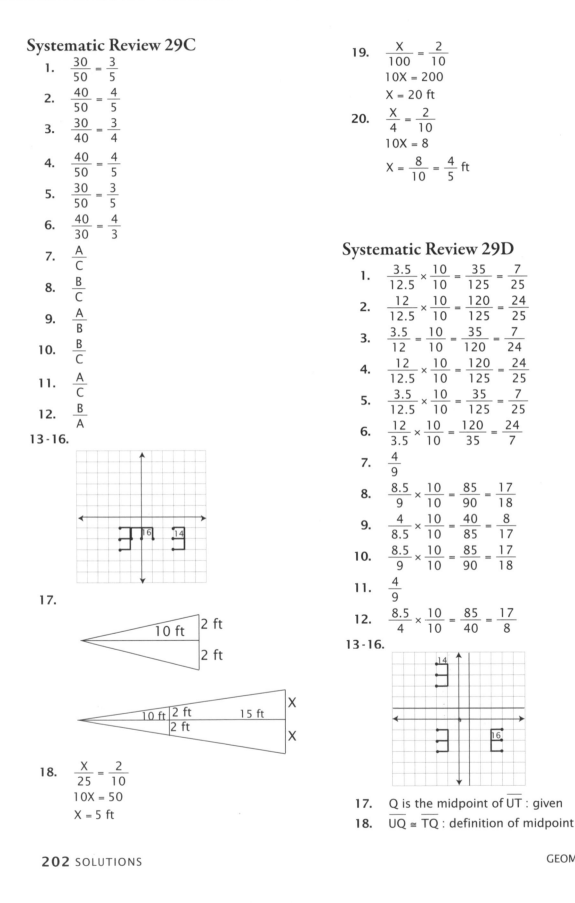

17.

18. $\dfrac{X}{25} = \dfrac{2}{10}$

$10X = 50$

$X = 5$ ft

19. $\dfrac{X}{100} = \dfrac{2}{10}$

$10X = 200$

$X = 20$ ft

20. $\dfrac{X}{4} = \dfrac{2}{10}$

$10X = 8$

$X = \dfrac{8}{10} = \dfrac{4}{5}$ ft

Systematic Review 29D

1. $\dfrac{3.5}{12.5} \times \dfrac{10}{10} = \dfrac{35}{125} = \dfrac{7}{25}$

2. $\dfrac{12}{12.5} \times \dfrac{10}{10} = \dfrac{120}{125} = \dfrac{24}{25}$

3. $\dfrac{3.5}{12} = \dfrac{10}{10} = \dfrac{35}{120} = \dfrac{7}{24}$

4. $\dfrac{12}{12.5} \times \dfrac{10}{10} = \dfrac{120}{125} = \dfrac{24}{25}$

5. $\dfrac{3.5}{12.5} \times \dfrac{10}{10} = \dfrac{35}{125} = \dfrac{7}{25}$

6. $\dfrac{12}{3.5} \times \dfrac{10}{10} = \dfrac{120}{35} = \dfrac{24}{7}$

7. $\dfrac{4}{9}$

8. $\dfrac{8.5}{9} \times \dfrac{10}{10} = \dfrac{85}{90} = \dfrac{17}{18}$

9. $\dfrac{4}{8.5} \times \dfrac{10}{10} = \dfrac{40}{85} = \dfrac{8}{17}$

10. $\dfrac{8.5}{9} \times \dfrac{10}{10} = \dfrac{85}{90} = \dfrac{17}{18}$

11. $\dfrac{4}{9}$

12. $\dfrac{8.5}{4} \times \dfrac{10}{10} = \dfrac{85}{40} = \dfrac{17}{8}$

13-16.

17. Q is the midpoint of $\overline{UT}$: given

18. $\overline{UQ} \cong \overline{TQ}$: definition of midpoint

19. $\angle UQS \cong \angle TQP$: vertical angles
20. $\triangle QSU \cong \triangle QPT$: AAS

Systematic Review 29E

1. $\dfrac{8}{12.8} \times \dfrac{10}{10} = \dfrac{80}{128} = \dfrac{5}{8}$

2. $\dfrac{10}{12.8} \times \dfrac{10}{10} = \dfrac{100}{128} = \dfrac{25}{32}$

3. $\dfrac{8}{10} = \dfrac{4}{5}$

4. $\dfrac{10}{12.8} \times \dfrac{10}{10} = \dfrac{100}{128} = \dfrac{25}{32}$

5. $\dfrac{8}{12.8} \times \dfrac{10}{10} = \dfrac{80}{128} = \dfrac{5}{8}$

6. $\dfrac{10}{8} = \dfrac{5}{4}$

7-10.

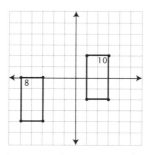

10. Appears the same as the original, because the figure is symmetrical around the X axis.

11. equiangular triangle, equilateral

12. long leg $\div \sqrt{3} = \dfrac{5\sqrt{6}}{\sqrt{3}} = \dfrac{5\sqrt{2}}{1} = 5\sqrt{2}$

13. short leg $\times 2 = \left(5\sqrt{2}\right)(2) = 10\sqrt{2}$

14. diameter

15. $40° + 85° = 125°$

16. $180° - 125° = 55°$

17. $180°$

18. $180° - 72° = 108°$
    (supplementary angles)

19. $2B + 3B = 180°$
    (supplementary angles)
    $5B = 180°$
    $B = 36°$
    $3B = X$ (alternate interior angles)
    $X = 3(36°) = 108°$
    $108° = X$

20. $6B + 14B = 180°$
    (supplementary angles)
    $20B = 180°$
    $B = 9°$
    $6B = X$ (alternate exterior angles)
    $X = 6(9°) = 54°$

Lesson Practice 30A

1. $\dfrac{6}{10} = \dfrac{3}{5}$

2. $\dfrac{8}{10} = \dfrac{4}{5}$

3. $\dfrac{6}{8} = \dfrac{3}{4}$

4. $\dfrac{5}{3}$

5. $\dfrac{5}{4}$

6. $\dfrac{4}{3}$

7. $\dfrac{24}{26} = \dfrac{12}{13}$

8. $\dfrac{10}{26} = \dfrac{5}{13}$

9. $\dfrac{24}{10} = \dfrac{12}{5}$

10. $\dfrac{13}{12}$

11. $\dfrac{13}{5}$

12. $\dfrac{5}{12}$

13. $\dfrac{4}{8.5} \times \dfrac{10}{10} = \dfrac{40}{85} = \dfrac{8}{17}$

14. $\dfrac{7.5}{8.5} \times \dfrac{10}{10} = \dfrac{75}{85} = \dfrac{15}{17}$

15. $\dfrac{4}{7.5} \times \dfrac{10}{10} = \dfrac{40}{75} = \dfrac{8}{15}$

16. $\dfrac{17}{8}$

17. $\dfrac{17}{15}$

18. $\dfrac{15}{8}$

19. 1

20. Pythagorean Theorem

Lesson Practice 30B

1. $\dfrac{3}{5}$

2. $\dfrac{4}{5}$

3. $\dfrac{3}{4}$

4. $\dfrac{5}{3}$

5. $\dfrac{5}{4}$

6. $\dfrac{4}{3}$

7. $\dfrac{15}{17}$

8. $\dfrac{8}{17}$

9. $\dfrac{15}{8}$

10. $\dfrac{17}{15}$

11. $\dfrac{17}{8}$

12. $\dfrac{8}{15}$

13. $\dfrac{4}{8} = \dfrac{1}{2}$

14. $\dfrac{5.7}{8} \times \dfrac{10}{10} = \dfrac{57}{80}$

15. $\dfrac{4}{5.7} \times \dfrac{10}{10} = \dfrac{40}{57}$

16. $\dfrac{2}{1}$

17. $\dfrac{80}{57}$

18. $\dfrac{57}{40}$

19. $\cos^2 \theta$

20. trig ratios

Systematic Review 30C

1. $\dfrac{12}{13.4} \times \dfrac{10}{10} = \dfrac{120}{134} = \dfrac{60}{67}$

2. $\dfrac{6}{13.4} \times \dfrac{10}{10} = \dfrac{60}{134} = \dfrac{30}{67}$

3. $\dfrac{12}{6} = \dfrac{2}{1}$

4. $\dfrac{67}{60}$

5. $\dfrac{67}{30}$

6. $\dfrac{1}{2}$

7. $\sin^2 \theta + \cos^2 \theta = 1$

8-11.

12. $\angle LPG \cong \angle RGP$:
    definition of parallelogram;
    alternate interior angles

13. $\angle LGP \cong \angle RPG$:
    definition of parallelogram;
    alternate interior angles

14. $\overline{PG} \cong \overline{PG}$: reflexive property

15. $\triangle PLG \cong \triangle GRP$: ASA

16. decagon

17. congruent

18. theorems

19. arc

20. $\sqrt{2}$

Systematic Review 30D

1. $\dfrac{13.6}{15} \times \dfrac{10}{10} = \dfrac{136}{150} = \dfrac{68}{75}$

2. $\dfrac{6.4}{15} \times \dfrac{10}{10} = \dfrac{64}{150} = \dfrac{32}{75}$

3. $\dfrac{13.6}{6.4} \times \dfrac{10}{10} = \dfrac{136}{64} = \dfrac{17}{8}$

4. $\dfrac{15}{13.6} \times \dfrac{10}{10} = \dfrac{150}{136} = \dfrac{75}{68}$

5. $\dfrac{75}{32}$ (inverse of #2)

6. $\dfrac{8}{17}$ (inverse of #3)

7. 1

8 - 9.

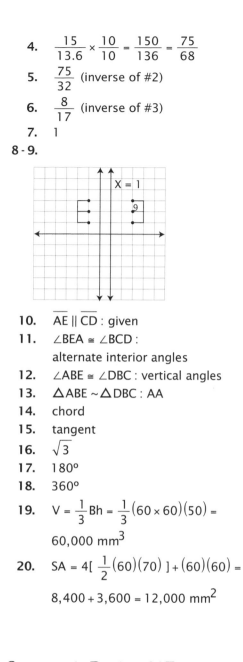

10. $\overline{AE} \parallel \overline{CD}$: given

11. $\angle BEA \cong \angle BCD$:
    alternate interior angles

12. $\angle ABE \cong \angle DBC$: vertical angles

13. $\triangle ABE \sim \triangle DBC$: AA

14. chord

15. tangent

16. $\sqrt{3}$

17. 180°

18. 360°

19. $V = \dfrac{1}{3}Bh = \dfrac{1}{3}(60 \times 60)(50) =$
    60,000 mm^3

20. $SA = 4[\dfrac{1}{2}(60)(70)] + (60)(60) =$
    $8,400 + 3,600 = 12,000$ mm^2

Systematic Review 30E

1. $\dfrac{24}{25}$

2. $\dfrac{7}{25}$

3. $\dfrac{24}{7}$

4. $\dfrac{25}{24}$

5. $\dfrac{25}{7}$

6. $\dfrac{7}{24}$

7. $\sin \theta$

8 - 9.

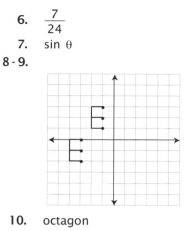

10. octagon

11. perpendicular

12. $r = \dfrac{1}{2}d = \dfrac{1}{2}(20) = 10$ ft

    $A = \pi r^2 \approx (3.14)(10)^2 = 314$ ft^2

13. express all measurements
    in the same unit:
    .75 ft × 12 = 9 in
    1.5 ft × 12 = 18 in
    $V = Bh = \dfrac{1}{2}(9)(5) \times 18 = 405$ in^3

14. SA = area of 2 ends plus area of
    2 sides plus area of bottom =
    $2(\dfrac{1}{2}(9)(5)) + 2((18)(7)) + (9)(18) =$
    $45 + 252 + 162 = 459$ in^2

15. $D = 2\pi r \approx 2(3.14)(4) = 25.12$ in

16. arc is $\dfrac{90°}{360°} = \dfrac{1}{4}$ of circle
    $\dfrac{1}{4} \times 25.12 = 6.28$ in

17. $\dfrac{(X+3)+(X+7)}{2} = 14$
    $\dfrac{2X+10}{2} = 14$
    $2X + 10 = 2(14)$
    $2X + 10 = 28$
    $2X = 18$
    $X = 9$ in

18. A = average base × height
   $= 14(X+2)$

$14(X+2) \Rightarrow 14((9)+2) = 14(11) = 154 \text{ in}^2$

19. $36 = (2A+3)+(2A+3)+(2A)$
   $36 = 6A+6$
   $30 = 6A$
   A = 5 ft

20. $42 = (2A+3)+(2A+3)+(2A)$
   $42 = 6A+6$
   $36 = 6A$
   A = 6 ft

Honors Solutions

Honors Lesson 1

1. Begin by putting x's to show that Tyler and Madison do not like tacos. That leaves Jeff as the one who has tacos as his favorite. Since you know Jeff's favorite, you can also put x's in Jeff's row, under ice cream and steak. We are told that Madison is allergic to anything made with milk, so we can put an x across from her name, under ice cream. Now we can see that Tyler is the only one who can have ice cream as his favorite, leaving Madison with steak.

	ice cream	tacos	steak
Jeff	X	yes	X
Tyler	yes	X	X
Madison	X	X	yes

2. We use similar reasoning for the rest of the problems. Remember that once you have a "yes" in any row or column, the rest of the possibilities in that row and in that column can be eliminated.

	black	brown	blonde
Mike	yes	X	X
Caitlyn	X	X	yes
Lisa	X	yes	X

3.

	reading	tennis	cooking	eating
George	X	X	yes	X
Celia	X	yes	X	X
Donna	yes	X	X	X
Adam	X	X	X	yes

4.

	spring	summer	autumn	winter
David	X	X	yes	X
Linda	X	X	X	yes
Shauna	yes	X	X	X
April	X	yes	X	X

Honors Lesson 2

1. $18 + 20 = 38$
   $38 - 30 = 8$ days had both

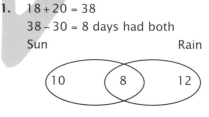

   $S \cap R = 8$
   $S \cup R = 30$

2. 1 (The twisted ring you started with is called a Mobius strip.)

3. 1st time : one long loop is created
   2nd time : two interlocked loops are created

4. $\left(5^2 + 5\right) \div 6 + 10 =$
   $\left(25 + 5\right) \div 6 + 10 =$
   $\left(30 \div 6\right) + 10 =$
   $5 + 10 = 15$

5. $\left(42 \div 7\right) + 6 - 1 =$
   $\left(6\right) + 6 - 1 =$
   $12 - 1 = 11$

Honors Lesson 3

1. 2
2. 4
3. $L \cap F = 3$
4. $L \cup F = 12$

5. 8
6. 3

7. $K \cup M - P = 12$

8. $K \cap M \cap P = 4$

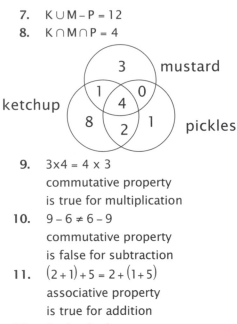

9. $3 \times 4 = 4 \times 3$
   commutative property
   is true for multiplication

10. $9 - 6 \neq 6 - 9$
    commutative property
    is false for subtraction

11. $(2 + 1) + 5 = 2 + (1 + 5)$
    associative property
    is true for addition

12. $2 \div 8 \neq 8 \div 2$
    commutative property
    is false for division

Honors Lesson 4

1. 45°

2. NNW

3. NNE

4. no, he should have corrected 67.5°

5. $5X - 6 = 2X + 18$
   $5X - 2X = 18 + 6$
   $3X = 24$
   $X = 8$

6. $2C + 10 = 43 - C$
   $3C = 33$
   $C = 11$

7. $(\$1.75 + D) + D = \$3.25$
   $2D + \$1.75 = \$3.25$
   $2D = \$1.50$
   $D = \$.75$
   $\$.75 + \$1.75 = \$2.50$
   Drink is $.75
   Sandwich is $2.50

8. let X = number of Isaac's customers
   2X = number of Aaron's customers
   $X + 2X = 105$
   $3X = 105$
   $X = 35$
   $2X = 70$
   Isaac has 35 customers
   Aaron has 70 customers

9. $X + 2X = 18$
   $3X = 18$
   $X = 6$ feet; $2X = 12$ feet

10. $A + (A + 20) = 144$
    $2A + 20 = 144$
    $2A = 124$
    $A = 62$ apples in one box
    $62 + 20 = 82$ apples in the other box

Honors Lesson 5

1.

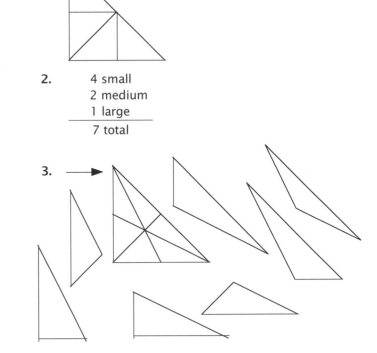

2. 4 small
   2 medium
   1 large
   ───────
   7 total

3.

4. 1 started with
2 that are half of first triangle
6 small
<u>7 overlapping</u>   (you may need to
16 total   draw these
                    separately to be
                    able to count each
                    one. See Above.)

5.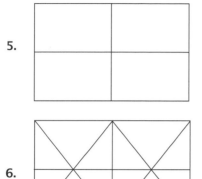

6.

Honors Lesson 6

1.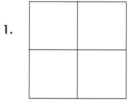

2.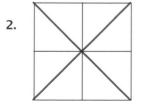

3. triangles, squares,
trapezoids, pentagons

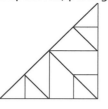

4. answers will vary

5. $P = 6X + .5(6)X$
$P = 6X + 3X$
$P = 9X$

6. $P = 9X$
$P = 9(8)$
$P = \$72$

Honors Lesson 7

1. Extend all segments
$\overline{AD} \parallel \overline{XY} \parallel \overline{BC}$
$\overline{AB} \parallel \overline{RS} \parallel \overline{DC}$
corresponding angles
are congruent

2. Yes; extend $\overline{DF}$ and $\overline{BC}$
these 2 line segments are
cut by transversal $\overline{AB}$
corresponding $\angle$'s ADF and
ABE are both 90°

3. extend $\overline{DC}$ to include point G
$m\angle A = 100°$
since $\overline{AB}$ and $\overline{DC}$ are parallel,
$m\angle GDA$ is 100°.
$m\angle EDF$ is 80°, since it is
supplementary to $\angle GDA$.
$m\angle DEF = 90°$ - definition
of perpendicular

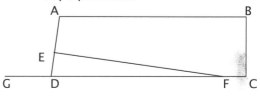

4. CAB = 90° (given)
    BAD = 45° –definition of bisector
    ADB = 90° - definition
    of perpendicular
    ABD = 45° - from information given
    DBE = 135° - supplementary angles
    all other corners work out
    the same way.

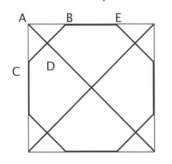

Honors Lesson 8

1. Look at the drawing below to
    see how the angles are labeled
    for easy reference.
    a and d are 25°
    definition of bisector

    p and o are 20°
    definition of bisector

    i and j are 45°
    definition of bisector

    Now look at triangle AEB. Its angles
    must add up to 180°. We know the
    measure of a and that of ABC. Add
    these together, and subtract the
    result from the total 180° that are
    in a triangle:
    $180 - (25 + 90) = 180 - 115 = 65°$
    l = 65°

Using similar reasoning, and looking
at triangles AEC, BFC, ABF, DBC and
ADC, we can find the following:
m = 115°
r = 95°   f = 85°
b = 110°.   g = 70°
Now we know two angles from each
of the smaller triangles. Armed with
this knowledge, and the fact that
there are 180° in a triangle, we can
find the remaining angles:
c = 45°   e = 70°
q = 65°   n = 45°
k = 70°   h = 65°
You can also use what you
know about vertical angles
and complementary angles
to find some of the angles.

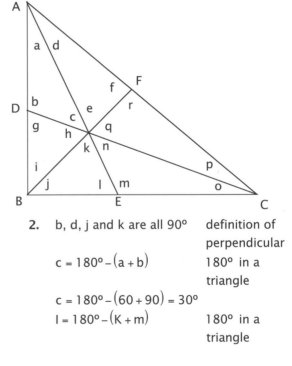

2. b, d, j and k are all 90°   definition of
                                 perpendicular
    $c = 180° - (a + b)$ 180° in a
                                 triangle
    $c = 180° - (60 + 90) = 30°$
    $l = 180° - (K + m)$ 180° in a
                                 triangle

$l = 180° - (90 + 30) = 60°$

$l + i = 90°$ Angle EGC is 90° because of the definition of perpendicular.

$60 + i = 90°$

$i = 30°$

$h = 180° - (i + j)$ 180Υin a triangle

$h = 180° - (30 + 90)$

$h = 60°$

$f + h = 180°$ Angle BEC is 180°

$f + 60 = 180°$

$f = 120°$

$c + g = 90°$ Angle AGE is 90° because of the definition of perpendicular.

$30 + g = 90°$

$g = 60°$

3. Use the same process for this one. Remember that you can also use what you know about vertical angles or complementary and supplementary angles as a shortcut.

● = 22.5°

○ = 45°

■ = 67.5°

□ = 90°

▲ = 112.5°

△ = 135°

Honors Lesson 9

1. large rectangle:

    15' 6" = 15.5 ft

    15.5 x 13 = 201.5 ft^2

    small rectangle:

    3 x 5 = 15 ft^2

    large trapezoid:

    $(9)(\frac{10+4}{2}) = (9)(\frac{14}{2}) = (9)(7) = 63$ ft

    small trapezoid:

    $(2)(\frac{4+8}{2}) = (2)(\frac{12}{2}) = (2)(6) = 12$ ft^2

    total:

    201.5 + 15 + 63 + 12 = 291.5 ft^2

2. It is necessary sometimes to add lines to the drawing to make it clearer. In figure 1a, dotted lines have been added to show how one end of the figure has been broken up. Since we know that the long measurement is 6.40 in and the space between the dotted lined is .80 in, we can see that the heights of the trapezoids add up to 5.60 in. Since we have been told that the top and bottom are the same, each trapezoid must have a height of 2.80 in.

Area of each trapezoid:

$$(2.8)(\frac{1.27+.80}{2}) = (2.8)(\frac{2.07}{2}) =$$

2.898 in^2

Since there are four trapezoids in all, we multiply by 4:

2.898 x 4 = 11.592 in^2

Rectangular center portion:

.80 in x 15 in = 12 in^2

Total:

12 + 11.592 = 23.592 in^2

3. area = $(a)(b)$ or ab (see figure 2)

4. area = $(2a)(2b)$ or 4ab (see figure 3)

5. area = $(na)(nb)$ or n^{2}ab (see figure 4)

6. area = n^{2}ab = $(5^2)(4)(5)$ = $(25)(20)$ = 500 ft^2

7. first triangle: a = $\frac{1}{2}$xy

second triangle: a = $\frac{1}{2}(2x)(2y)$ = 2xy

4 times $\frac{1}{2}$ = 2, so new area is four times as great.

8. first square: $(x)(x) = x^2$

second square: $(x^2)(x^2) = x^4$

figure 1a

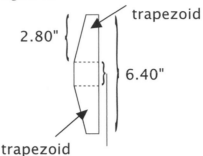

trapezoid

2.80"

6.40"

trapezoid

figure 1b (shows a different way of finding the area)

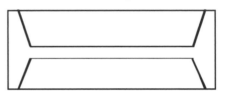

Area of large rectangle 15 x 6.4 = 96 in^2

One trapezoid	long base
	15 – $(2 \times .8)$ = 13.4 in^2
	short base
	15 – (2×1.27) =
	12.46 in^2
	height $(6.4 - .8) \div$
	2 = 2.8 in^2
	Area of one trapezoid
	= 36.204 in^2
Both trapezoids	2 x 36.204 =
	72.408 in^2
Area of figure	96 – 72.408 =
	23.592 in^2

figure 2

a [ ]
b

figure 3

2a [ ]
2b

figure 4

na

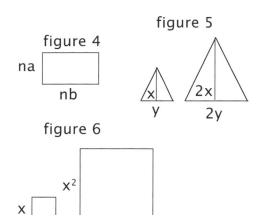

figure 5

nb

figure 6

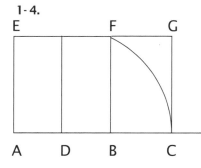

Honors Lesson 10

1-4.

5. your answer should be close to 0.61803.

6. See illustration above.
   The ratio should be close to what you got in #5.

7-8.

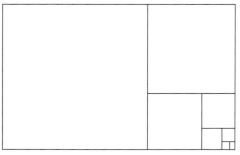

Honors Lesson 11

1.

	green, buttons	green, zipper	red, zipper	blue, buttons
Chris	yes	x	x	x
Douglas	x	yes	x	x
Ashley	x	x	x	yes
Naomi	x	x	yes	x

2.

	planning games	refresh-ments	place for party	birthday guest
Sam	x	x	yes	x
Jason	x	x	x	yes
Shane	yes	x	x	x
Troy	x	yes	x	x

3.

	train	boat	airplane	car
Janelle	yes	x	x	x
Walter	x	x	x	yes
Julie	x	yes	x	x
Jared	x	x	yes	x

4.

	hot dog	pizza	chicken soup	tossed salad
Molly	yes	x	x	x
Tina	x	x	x	yes
Logan	x	x	yes	x
Sam	x	yes	x	x

5. Answers will vary.

Honors Lesson 12

1.

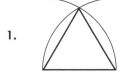

2. 60°

3. Since the sections are all equal, the center angles are all the same. 360° ÷ 8 = 45°

4 - 8.

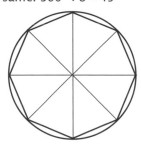

9. In #3, you divided 360° by 8 to find that each small triangle has a central angle of 45°. Since a hexagon has six sides, you want to construct six triangles inside the circle. 360° ÷ 6 = 60°

In #1, you learned how to construct an equilateral triangle with each angle equal to 60°. After drawing a circle and one diameter, use the same procedure to construct equilateral triangles inside your circle, using a radius of the circle as your starting point each time. After you have constructed four triangles, connect their points, and you will have an inscribed regular hexagon.

10 - 12.

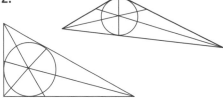

Honors Lesson 13

1. See illustration.

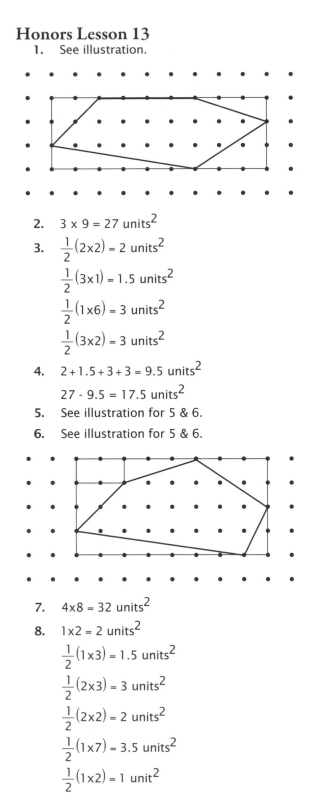

2. $3 \times 9 = 27$ units^2

3. $\frac{1}{2}(2 \times 2) = 2$ units^2

$\frac{1}{2}(3 \times 1) = 1.5$ units^2

$\frac{1}{2}(1 \times 6) = 3$ units^2

$\frac{1}{2}(3 \times 2) = 3$ units^2

4. $2 + 1.5 + 3 + 3 = 9.5$ units^2

$27 - 9.5 = 17.5$ units^2

5. See illustration for 5 & 6.

6. See illustration for 5 & 6.

7. $4 \times 8 = 32$ units^2

8. $1 \times 2 = 2$ units^2

$\frac{1}{2}(1 \times 3) = 1.5$ units^2

$\frac{1}{2}(2 \times 3) = 3$ units^2

$\frac{1}{2}(2 \times 2) = 2$ units^2

$\frac{1}{2}(1 \times 7) = 3.5$ units^2

$\frac{1}{2}(1 \times 2) = 1$ unit^2

9. $2 + 1.5 + 3 + 2 + 3.5 + 1 = 13$ units^2

    $32 - 13 = 19$ units^2

10. See illustration.

    $10 \times 5 = 50$ units^2

    $\frac{1}{2}(5 \times 2) = 5$ units^2

    $1 \times 5 = 5$ units^2

    $\frac{1}{2}(1 \times 5) = 2.5$ units^2

    $\frac{1}{2}(4 \times 1) = 2$ units^2

    $1 \times 1 = 1$ unit^2

    $\frac{1}{2}(6 \times 1) = 3$ units^2

    $\frac{1}{2}(3 \times 4) = 6$ units^2

    $5 + 5 + 2.5 + 2 + 1 + 3 + 6 = 24.5$ units^2

    $50 - 24.5 = 25.5$ units^2

11. See illustration.

    $4 \times 8 = 32$ units^2

    $\frac{1}{2}(1 \times 1) = .5$ units^2

    $\frac{1}{2}(3 \times 1) = 1.5$ units^2

    $1 \times 1 = 1$ unit^2

    $\frac{1}{2}(1 \times 2) = 1$ unit^2

    $\frac{1}{2}(1 \times 2) = 1$ unit^2

    $1 \times 1 = 1$ unit^2

    $\frac{1}{2}(5 \times 1) = 2.5$ units^2

    $\frac{1}{2}(2 \times 1) = 1$ unit^2

    $.5 + 1.5 + 1 + 1 + 1 + 1 + 2.5 + 1$

    $= 9.5$ units^2

    $32 - 9.5 = 22.5$ units^2

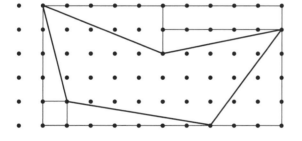
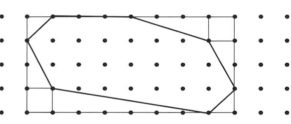

Honors Lesson 14

1. $\frac{1}{2}(3 \times 4) = \frac{1}{2}(12) = 6$ units^2

2. $A = \sqrt{s(s-a)(s-b)(s-c)}$

    $A = \sqrt{6(6-3)(6-4)(6-5)}$

    $A = \sqrt{6(3)(2)(1)}$

    $A = \sqrt{36}$

    $A = 6$ units^2

    yes

3. $A = \sqrt{16(16-7)(16-10)(16-15)}$

   $A = \sqrt{16(9)(6)(1)}$

   $A = \sqrt{864}$

   $A = 29.39$ units^2

4. $A = \sqrt{52(52-36)(52-28)(52-40)}$

   $A = \sqrt{52(16)(24)(12)}$

   $A = \sqrt{239,616}$

   $A = 489.51$ units^2

5. $V = \pi r^2 h$

   $V = 3.14(2)^2(10)$

   $V = 3.14(4)(10)$

   $V = 125.6$ in^3

6. $V = \pi r^2 h$

   $V = 3.14(1)^2(10)$

   $V = 3.14(1)(10)$

   $V = 31.4$ in^3

   It is $\frac{1}{4}$ the first one

7. $V = \pi r^2 h$

   $V = 3.14(2)^2(5)$

   $V = 3.14(4)(5)$

   $V = 62.8$ in^3

   It is half the first one.

8. $V = \pi r^2 h$

   $V = 3.14(4)^2(10)$

   $V = 3.14(16)(10)$

   $V = 502.4$ in^3

   It is four times the first one.

9. $V = \pi r^2 h$

   $V = 3.14(2)^2(20)$

   $V = 3.14(4)(20)$

   $V = 251.2$ cu in^3

   It is two times the first one

10. When the height is doubled, the volume is doubled. When the height is halved, the volume is halved. When the radius is doubled, the volume increases by a factor of 4. When the radius is halved, the volume decreases by a factor of 4.
    The student may use his own words to express this.

11. Answers will vary.

12. Take the formula, and multiply both sides by 2:

    $V = \pi r^2 h$

    $2V = 2\pi r^2 h$

    Now rearrange the factors:

    $V = \pi r^2 h$

    $2V = \pi r^2 2h$

    Take the formula, and multiply both sides by 4:

    $V = \pi r^2 h$

    $4V = 4\pi r^2 h$

    Rewrite the 4 on the right side as 2^2:

    $4V = 2^2 \pi r^2 h$

    Rearrange the factors:

    $4V = \pi 2^2 r^2 h$

    $4V = \pi (2r)^2 h$

    There is more than one way to set this up. As long as you show the same results as by experimentation, the answer is correct.

Honors Lesson 15

1. $3 \times 3 \times 3 = 27$ ft

2. $12 \times 12 \times 12 = 1{,}728$ in^3

3. $8 \times 4 \times 2 = 64$ in^3
$64 \times .3 = 19.2$ lb

4. 64 in$^3 \div 1{,}728 = .037$ ft^3
$.037 \times 1200 = 44.4$ lbs
You could probably lift it,
but it would be much heavier
than expected.

5. First find what the volume would
be if it were solid:
$V = \pi r^2 h$
$V = 3.14(.5)^2(12)$
$V = 9.42$ in^3
Now find the volume inside
the pipe:
$V = \pi r^2 h$
$V = 3.14(.25)^2(12)$
$V = 2.355$ in^3
Then find the difference:
$9.42 - 2.355 = 7.065$ in^3

6. $7.065 \times .26 = 1.8369$ lb

7. $V = \frac{4}{3}\pi r^3$
$V = \frac{4}{3}(3.14)(.25)^3$
$V = .07$ in^3 (rounded)
$.07 \times .3 \approx .02$ pounds for
one bearing
$25 \div .02 = 1{,}250$ bearings
Because we rounded some
numbers, the actual number
of bearings in the box may be
slightly different. Keep in mind
that the starting weight was
rounded to a whole number.
Our answer is close enough to
be helpful in a real life situation,
where someone wants to know
approximately how many bearings
are available without counting.

8. The side view is a trapezoid,
and the volume of the water
is the area of the trapezoid
times the width of the pool:
$A = \frac{3+10}{2}(40)$
$A = 6.5(40)$
$A = 260$ ft^2
$V = 260(20)$
$V = 5{,}200$ ft^3

9. Volume of the sphere:
$V = \frac{4}{3}(3.14)(1)^3$ units^3
$V = 4.19$
Volume of the cube:
$V = 2 \times 2 \times 2 = 8$ units^3
$8 - 4.19 = 3.81$ units^3

10. Volume of the cylinder:

$V = 3.14 \, (1)^2 (2)$

$V = 6.28 \text{ units}^3$

Volume of the sphere from #9:

4.19 units^3

$6.28 - 4.19 = 2.09 \text{ units}^3$

Note: You may use the fractional value of π if it seems more convenient.

Honors Lesson 16

1. $(r)\pi r = \pi r^2$
2. $A = LW + LW + LH + LH + WH + WH$
   $= 2LW + 2LH + 2WH$
   $= 2(LW + LH + WH)$
3. $2(s^2 + s^2 + s^2) = 2(3s^2) = 6s^2$
4. $V = 3(11)(3) = 99 \text{ ft}^3$
   $SA = 2(3 \times 11) + 2(3 \times 3) + 2(11 \times 3)$
   $= 2(33) + 2(9) + 2(33)$
   $= 66 + 18 + 66$
   $= 150 \text{ ft}^2$
5. $150 \text{ ft}^2 \div 6 \text{ faces} = 25 \text{ ft}^2 \text{ per face}$
   $\sqrt{25} = 5 \text{ ft}$
   The new bin is 5 x 5 x 5.
6. The cube-shaped one holds more.
   $125 - 99 = 26 \text{ ft}^3 \text{ difference.}$

Honors Lesson 17

1. $V = \pi r^2 h$
   $V = 3.14(2)^2(4)$
   $V = 50.24 \text{ ft}^3$

2. $V = \frac{4}{3}\pi r^3$
   $V = \frac{4}{3}(3.14)(2)^3$
   $V = 33.49 \text{ ft}^3 (\text{rounded})$
3. $V = 3.14(3)^2(6)$
   $V = 169.56 \text{ units}^3$
4. $V = \frac{4}{3}(3.14)(3)^3$
   $V = 113.04 \text{ units}^3 \text{ (rounded)}$
5. $V = 3.14(1)^2(2)$
   $V = 6.28 \text{ units}^3$
6. $V = \frac{4}{3}(3.14)(1)^3$
   $V = 4.19 \text{ units}^3 \text{ (rounded)}$
7. $\dfrac{33.49}{50.24} \approx .67 \qquad \dfrac{113.04}{169.56} \approx .67$
   $\dfrac{4.19}{6.28} \approx .67$
8. $\dfrac{2}{3}$
9. $A = 2\pi r^2 + 2\pi rh$
   $A = 2(3.14)(3)^2 + 2(3.14)(3)(6)$
   $A = 56.52 + 113.04 = 169.56 \text{ units}^2$
10. $A = 4(3.14)(3)^2$
    $A = 113.04 \text{ units}^2$
11. $\dfrac{113.04}{169.56} \approx \dfrac{2}{3}$
12. The surface area and volume of a sphere appear to be $\dfrac{2}{3}$ of the surface area and volume of a cylinder with the same dimensions. (Archimedes proved that this is the case.)

Honors Lesson 18

1. 4,003 mi
2. 90°; a tangent to a circle is perpendicular to the diameter

3. $L^2 + 4{,}000^2 = 4{,}003^2$

$L^2 = 4{,}003^2 - 4{,}000^2$

$L^2 = 16{,}024{,}009 - 16{,}000{,}000$

$L^2 = 24{,}009$

$L = \sqrt{24{,}009} \approx 155 \text{ mi}$

4. $29{,}035 \div 5{,}280 \approx 5 \text{ mi}$

5. $L^2 + 4{,}000^2 = 4{,}005^2$

$L^2 + 16{,}000{,}000 = 16{,}040{,}025$

$L^2 = 16{,}040{,}025 - 16{,}000{,}000$

$L^2 = 40{,}025$

$L = \sqrt{40{,}025} \approx 200 \text{ mi}$

6. $555 \div 5{,}280 \approx .1$

$L^2 + 4{,}000^2 = 4{,}000.1^2$

$L^2 + 16{,}000{,}000 = 16{,}000{,}800.01$

$L^2 = 16{,}000{,}800.01 - 16{,}000{,}000$

$L^2 = 800.01$

$L = \sqrt{800.01} \approx 28.3 \text{ mi}$

7. $150^2 + 4{,}000^2 = (X + 4{,}000)^2$

8. $X^2 + 8{,}000X + 16{,}000{,}000$

9. $22{,}500 + 16{,}000{,}000$

$= X^2 + 8{,}000X + 16{,}000{,}000$

$22{,}500 = X^2 + 8{,}000X$

$0 = X^2 + 8{,}000X - 22{,}500$

or $X^2 + 8{,}000X - 22{,}500 = 0$

10. $8{,}000X = 22{,}500$

$X = 22{,}500 \div 8{,}000$

$X \approx 2.8 \text{ mi}$

Honors Lesson 19

1. V = area of base x altitude

$V = (4 \cdot 4)(8)$

$V = 128 \text{ in}^3$

2. $SA = 2(4\text{x}4) + 2(4\text{x}8) + 2(4\text{x}8)$

$SA = 2(16) + 2(32) + 2(32)$

$SA = 32 + 64 + 64$

$SA = 160 \text{ in}^2$

3. V = area of base x altitude

$V = \frac{1}{2}(3\text{x}4)\text{x}10$

$Y = 60 \text{ ft}^3$

4. $SA = (2)\frac{1}{2}(3\text{x}4) + (3\text{x}10) +$

$(4\text{x}10) + (5\text{x}10)$

$SA = 12 + 30 + 40 + 50$

$SA = 132 \text{ ft}^2$

5. Think of the wire as a long, skinny cylinder.

$1 \text{ ft}^3 = 12\text{x}12\text{x}12 = 1{,}728 \text{ in}^3$

Volume of wire = area of base x length

$1{,}728 = (3.14 \text{ x } .1^2) \text{x } L$

$1{,}728 = .0314L$

$55{,}031.8 \text{ in} \approx L$

$55{,}031.8 \div 12 \approx 4{,}586 \text{ ft}$

6. $A = LW$ Let L = the circumference and W = the height of the cylinder.

Diameter = 9, so $L = 3.14(9)$

$28.26 \text{ in} \approx L$ This is one dimension of the rectangle and the circumference of the cylinder.

$625 = 28.26W$

$22.12 \text{ in} = W$ This is the other dimension of the rectangle and the height of the cylinder.

V = area of base x height

$V = 3.14(9 \div 2)^2 \text{ x } 22.12$

$V = 3.14(4.5)^2 \text{ x } 22.12$

$V \approx 1{,}406.5 \text{ in}^3$

7. Cylinder will be 4 in high and 4 in^3 in diameter. Area of one circular end = $3.14(2)^2 = 12.56$ in^2

area of side = $3.14(4) \times 4 = 50.24$ in^2

$50.24 + 12.56 + 12.56 = 75.36$ in^2

You also could have used what you learned in lesson 17 to find the surface area of the cylinder. First find the surface area of the sphere, and then multiply by $\frac{3}{2}$. (See below for an alternative solution.)

7. alternative solution

SA of sphere = $4(3.14)(2)^2 =$ 50.24 in^2

$\frac{3}{2}$ or $1.5(50.24) = 75.36$ in^2

8. $A = 2(4 \times 4) + 2(4 \times 4) + 2(4 \times 4)$

$A = 32 + 32 + 32 = 96$ in^2

The cylinder uses less cardboard. (However, there will be odd-shaped, possibly unuseable pieces left over.)

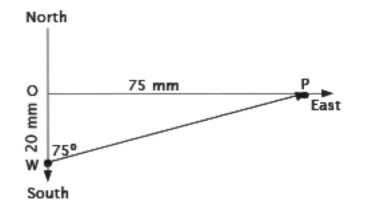

Honors Lesson 20

1. $300 \div 150 = 2$ hours
2. answers may vary
   The wind blew him off course.
3. $30 \div 2 = 15$ mm
4. $150 \div 2 = 75$ mm
5. $\angle OWP = 80°$
6. $\angle OWP = 75°$
   See drawing.
   Your answers to #5 and #6 may vary slightly depending how carefully you drew and measured.

Honors Lesson 21

1. πy^2
2. $A = \pi x^2 - \pi y^2$
3. $y^2 + z^2 = x^2$
   $z^2 = x^2 - y^2$
4. $A = \pi\left(x^2 - y^2\right)$
5. $A = \pi\left(z^2\right)$
6. $A = \pi\left(z^2\right)$
   $A = \pi\left(\frac{10}{2}\right)^2$
   $A = \pi(5)^2$
   $A = 3.14 \times 25$
   $A = 78.5$ in^2
7. $A = 3.14(4)^2$
   $A = 3.14 \times 16 = 50.24$ in^2
8. $A = L \times W$
   $50.24 = L \times .007$
   $50.24 \div .007 \approx 7,177$ in
9. $7,177 \div 2 \approx 3,589$ tickets (rounded to the nearest whole number)

Honors Lesson 22
1. This bird is red.
2. ∠A is congruent to ∠B.
3. I get 100% on my math test.
4. This triangle has two congruent sides.

Honors Lesson 23
1. If I get burned, I touched the hot stove. Not necessarily true.
2. If two line segments are congruent, they have equal length.
   True.
3. If a bird is red, it is a cardinal.
   Not necessarily true.
4. If the leg squared plus the leg squared equals the hypotenuse squared, the triangle is a right triangle.
   True.
5. If my plants wilt, I stop watering them.
   Not true if I am sensible!

Honors Lesson 24
1. 50°; the measure of an inscribed angle is half the measure of the intercepted arc.
2. 130°; $180° - 50°$
3. 50°; same reason as #1
4. 80°; $180° - (50° + 50°)$
5. 160°; $360° - (100° + 100°)$
6. 80°; vertical angles
7. 85°; $180° - 95°$
8. 15°; $180 - (80° + 85°)$
   checking results with remote interior angles: $80° + 15° = 95°$

9. 80°; angle 1 and the 70° angle next to it put together form an angle that is the alternate interior angle to the 150° angle at the top left.
   $150° - 70° = 80°$
10. 70°; alternate interior angles
11. 30°; $180° - (70° + 80°)$
12. 30°; alternate interior angles

Honors Lesson 25
1.

Statements	Reasons
$\overline{AF} \cong \overline{EF}$	Given
∠1 ≅ ∠2	Given
$\overline{CF} \cong \overline{CF}$	Reflexive
△CEF ≅ △CAF	SAS
$\overline{CE} \cong \overline{CA}$	Corresponding parts of congruent triangles
△CDE ≅ △CBA	SSS
∠CDE ≅ ∠CBA	Corresponding parts of congruent triangles

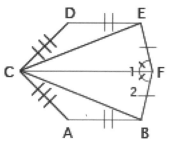

2.

Statements	Reasons
$\overline{TU} \cong \overline{RQ}$	Given
$\angle TUV \cong \angle RQV$	Given
$\overline{UV} \cong \overline{QV}$	Given
$\triangle TUV \cong \triangle RQV$	SAS
$\overline{TV} \cong \overline{RV}$	CPCTRC
$\overline{ST} \cong \overline{SR}$	Given
$\overline{SV} \cong \overline{SV}$	Reflexive
$\triangle TSV \cong \triangle RSV$	SSS
$\angle TSV \cong \angle RSV$	CPCTRC

3.

Statements	Reasons
$\overline{FE} \cong \overline{GH}$	Given
$\overline{FH} \cong \overline{GE}$	Given
$\overline{EH} \cong \overline{EH}$	Reflexive
$\triangle FEH \cong \triangle GHE$	SSS

Honors Lesson 26

1.

Statements	Reasons
$\overline{AB} \cong \overline{AC}$	Given
$\angle ARB \cong \angle AQC$	Perpendicular
$\angle BAR \cong \angle CAQ$	Reflexive
$\triangle BAR \cong \triangle CAQ$	AAS or HA
$\overline{CQ} \cong \overline{BR}$	CPCTRC

2.

Statements	Reasons
$\overline{XB} \cong \overline{YB}$	Definition of bisector
$\angle XBA \cong \angle YBA$	Definiton of Perpendicular
$\overline{BA} \cong \overline{BA}$	Reflexive
$\triangle XBA \cong \triangle YBA$	SAS or LL
$\overline{XA} \cong \overline{YA}$	CPCTRC

3.

Statements	Reasons
$\overline{EF} \cong \overline{GF}$	From proof above
$\overline{EX} \cong \overline{GX}$	Definition of Bisector
$\overline{FX} \cong \overline{FX}$	Reflexive
$\triangle EFX \cong \triangle GFX$	SSS or HL
$\angle EXH \cong \angle GXH$	Definition of Perpendicular
$\overline{HX} \cong \overline{HX}$	Reflexive
$\triangle EHX \cong \triangle GHX$	SAS or LL
$\overline{EH} \cong \overline{GH}$	CPCTRC

Honors Lesson 27

1.

Statements	Reasons
$\overline{XC} \cong \overline{YC}$	Radius of a circle
$\angle PYC \cong \angle PXC$	A tangent of a circle is perpendicular to the radius at that point.
$\overline{PC} \cong \overline{PC}$	Reflexive
$\triangle PYC \cong \triangle PXC$	HL
$\overline{PX} \cong \overline{PY}$	CPCTRC

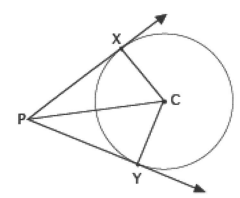

2.

Statements	Reasons
$\overline{DE} \perp \overline{AB}$	Given
$\overline{AC} \cong \overline{BC}$	Radius of a circle
$\overline{FC} \cong \overline{FC}$	Reflexive
$\triangle FCA \cong \triangle FCB$	HL
$\angle ACE \cong \angle BCE$	CPCTRC
$\overparen{AE} \cong \overparen{BE}$	Property of central angle

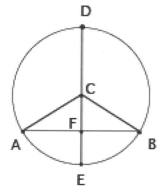

3.

Statements	Reasons
$\overline{OP} \cong \overline{LM}$	Given
$\overline{OC} \cong \overline{LC}$	Radius of a circle
$\overline{PC} \cong \overline{MC}$	Radius of a circle
$\triangle CPO \cong \triangle CML$	SSS
$\overline{OX} \cong \overline{LY}$	Definition of Bisector
$\triangle OCX \cong \triangle LCY$	HL
$\overline{XC} \cong \overline{YC}$	CPCTRC

Honors Lesson 28

1. $67 = (\frac{1}{2})X$

   $134° = X$

   $50(\frac{1}{2}) = Y$

   $100° = Y$

   $180 - (50 + 67) = (\frac{1}{2})Z$

   $63 = (\frac{1}{2})Z$

   $126° = Z$

2. $B = 180 - 77 = 103°$

   $A = 180 - 84 = 96°$

   $C = 2 \times 77 = 154°$

3. $m\overparen{QR} = 2(63°) = 126°$

   $m\angle QCR = m\overparen{QR} = 126°$

4. $m\angle AEC = \dfrac{40° + 30°}{2} = \dfrac{70°}{2} = 35°$

   $m\angle BED = 35°$

5. $m\angle KPL = \dfrac{116° - 36°}{2} = \dfrac{80°}{2} = 40°$

Honors Lesson 29

1.

angle	tan
10°	.18
15°	.268
30°	.58
45°	1
60°	1.73

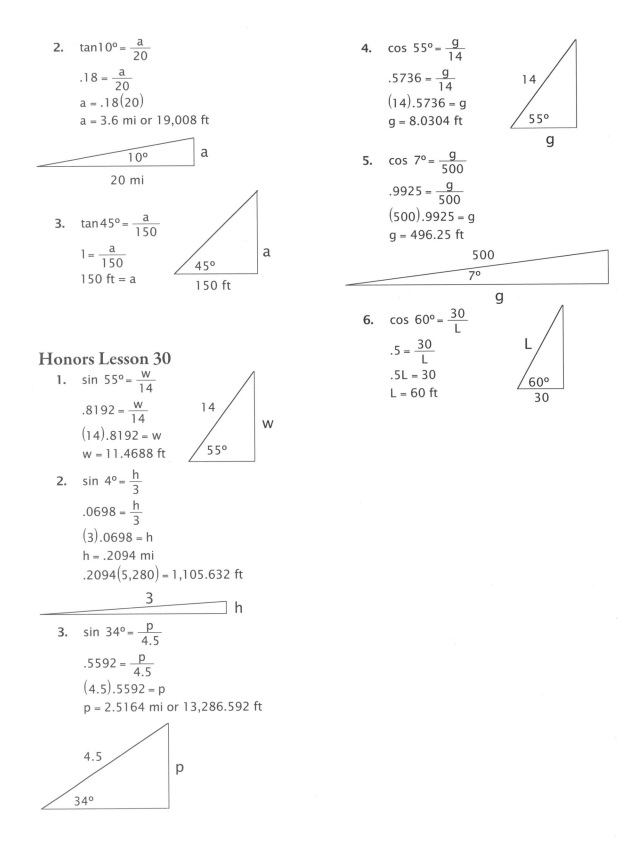

2. $\tan 10° = \dfrac{a}{20}$

   $.18 = \dfrac{a}{20}$
   $a = .18(20)$
   $a = 3.6$ mi or $19,008$ ft

3. $\tan 45° = \dfrac{a}{150}$

   $1 = \dfrac{a}{150}$
   150 ft $= a$

Honors Lesson 30

1. $\sin 55° = \dfrac{w}{14}$

   $.8192 = \dfrac{w}{14}$
   $(14).8192 = w$
   $w = 11.4688$ ft

2. $\sin 4° = \dfrac{h}{3}$

   $.0698 = \dfrac{h}{3}$
   $(3).0698 = h$
   $h = .2094$ mi
   $.2094(5,280) = 1,105.632$ ft

3. $\sin 34° = \dfrac{p}{4.5}$

   $.5592 = \dfrac{p}{4.5}$
   $(4.5).5592 = p$
   $p = 2.5164$ mi or $13,286.592$ ft

4. $\cos 55° = \dfrac{g}{14}$

   $.5736 = \dfrac{g}{14}$
   $(14).5736 = g$
   $g = 8.0304$ ft

5. $\cos 7° = \dfrac{g}{500}$

   $.9925 = \dfrac{g}{500}$
   $(500).9925 = g$
   $g = 496.25$ ft

6. $\cos 60° = \dfrac{30}{L}$

   $.5 = \dfrac{30}{L}$
   $.5L = 30$
   $L = 60$ ft

Test Solutions

Test 1

1. B : no dimensions
2. E : line
3. A : point
4. A : length
5. B : is infinite
6. C : point E
7. A : point Q
8. D : ° (infinite)
9. E : line segment $\overline{AB}$
10. B : ray CE
11. C : point
12. C : equal
13. A : similar
14. B : congruent
15. A : ray

Test 2

1. E : coplanar
2. D : line
3. C : length and width
4. C : two
5. A : intersection
6. B : union
7. B : infinite
8. C : ∅ (null or empty set)
9. A : ∩ (intersection)
10. E : ∪ (union)
11. B : line CD
12. C : line AH
13. E : points F, D, E
14. C : line EG
15. A : point E

Test 3

1. B : point H
2. B : point H
3. C : point J
4. A : point H
5. E : ∠AHC
6. B : protractor
7. C : degrees
8. B : ∠SRT
9. E : the measure of angle two
10. B : 45°
11. A : 90°
12. D : 80°
13. E : 130°
14. A : 25°
15. E : 160°

Test 4

1. D : ∠BGE
2. C : ∠CGE
3. A : ∠AGC
4. C : reflex
5. E : straight
6. B : point S
7. B : 90°
8. E : can't tell from information given; it may be anything between 90° and 180°
9. A : acute
10. D : 90°
11. C : ∠RNP
12. E : 30°+45° = 75°
13. B : 90°−50° = 40°
14. B : Of angles given, only obtuse ones are ∠MNQ and ∠PNS. ∠PNS is the smaller of the two.
15. C : m∠MNR = 60°+30° = 90° 90°+91° = 181° Reflex angles are > 180°

TEST 1 - TEST 4 225

Test 5

1. A : parallel
2. B : perpendicular
3. E : perpendicular
4. B : bisector
5. A : $AF = FB$
6. D : $\overline{DA}$ and $\overline{GF}$
7. C : I, II and IV are true
8. B : $90° \div 2 = 45°$
9. B : $90° \div 2 = 45°$
10. C : $\perp$
11. A : $\parallel$
12. A : This is the converse of the original statement.
13. C : I and III: straightedge and compa
14. D : at the vertex
15. C : perpendicular lines are not parallel

Test 6

1. E : supplementary
2. C : congruent
3. B : $90° - 35° = 55°$
4. C : $180° - 40° = 140°$
5. E : $20° + 70° = 90°$, so they are complementary
6. B : $\angle 2$ and $\angle 5$
7. A : $90°$, because line SV $\perp$ line WT
8. E : can't tell from information given
9. D : $\angle 1$
10. A : $180°$ They combine to form a straight angle.
11. C : vertical angles
12. D : We don't know the measures of $\angle 4$ and $\angle 5$, so sum cannot be determined.
13. A : $\overleftrightarrow{FC}$ is a straight line, so $\angle 1$ would be included to make $180°$.

14. D : The measures of these angles are not given: looking the same is not sufficient.
15. A : $90° + 90° < 185°$

Test 7

1. D : $\angle 7$
2. C : $180° - 80° = 100°$
3. E : Alternate interior angles are congruent.
4. B : $\angle 2$
5. D : alternate exterior angles
6. E : $\angle$'s 1, 2, 4, 5, 6, 7 and 8
7. C : $65°$; vertical angles
8. D : vertical angles
9. E : supplementary angles
10. E : can't tell: rules for alternate exterior angles apply only for parallel lines
11. C : If the sum of two angles is $180°$, they are supplementary.
12. A : parallel lines
13. D : $45°$
14. D : 8: four for each intersection
15. B : congruent

Test 8

1. E : I, II and V
2. C : All squares have 4 right angles and opposite sides that are congruent, so they are rectangles.
3. D : Some trapezoids have 1 right angle, but they need not have any.
4. E : length of each side
5. A : quadrilateral
6. D : $180°$
7. D : square
8. B : rhombus

9. A : 360°
10. B : trapezoid
11. A : 5 + 7 + 9 + 3 = 24 in
12. C : 9 + 10 + 15 = 34 m
13. D : unlabeled horizontal side
 has a length of 8.5 – 4 = 4.5 in
 P = 4 + 3 + 4.5 + 2 + 8.5 + 5 = 27 in
14. B : P = 4(11) = 44 cm
15. E : P = 2(25) + 2(15) = 50 + 30 = 80 ft

Test 9
1. B : height
2. E : perpendicular to the base
3. B : divide by two
4. D : find the average base
5. A : 90°, because they are perpendicular
6. C : 100 ft^2: area is always in square units
7. E : not enough information; need to know both bases
8. D : $A = \frac{1}{2}bh = \frac{1}{2}(8)(4) = 16$ m^2
9. C : $A = bh = (15)(3) = 45$ units^2
10. E : not enough information; perpendicular height is needed
11. A : $A = \frac{5+9}{2}(3.5) = 24.5$ in^2
12. A : $A = \frac{1}{2}bh = \frac{1}{2}(15)(6) = 45$ m^2
13. C : $A = (4)(3) + (2)(8.5) = 12 + 17 = 29$ ft^2
14. E : $A = bh = (11)(10) = 110$ cm^2
 all 4 sides of a rhombus are congruent
15. A : $A = bh = (25)(15) = 375$ ft^2

Test 10
1. D : obtuse
2. C : isosceles
3. B : acute
4. C : If it has a 90° angle, the remaining two angles must add to 90°. 90° – 28° = 62°
5. B : scalene
6. E : impossible to draw 61° + 62° + 61° = 184°
7. B : acute and equilateral
8. C : 6 is the smallest number among the choices which, when added to 7 yields a result greater than 12.
9. E : impossible to draw 2 + 2 = 4; 4 < 5
10. A : isosceles and right
11. C : equilateral, because the third angle must also be 60°
12. D : right 180° – (74° + 16°) = 180° – 90° = 90°
13. A : isosceles and acute
14. B : equilateral
15. A : 34° + 73° + 73° = 180°

Unit Test I
I
1. intersection
2. congruent
3. empty or null
4. triangle
5. supplementary
6. reflex
7. bisector
8. angle

II
1. check with protractor
2. check with protractor: smaller angles should each measure 40°

III

1. a triangle with two equal sides
2. a triangle with no equal sides

IV

$P = 4 + 9 + 5 + 11 = 29$ ft

$A = \dfrac{11 + 9}{2}(3) = 10(3) = 30$ ft^2

V

1. $110°$: vertical angles
2. $m\angle 11 = m\angle 3 = 110°$
   corresponding angles
   $m\angle 9 = 180° - m\angle 11 =$
   $180° - 110° = 70°$
   supplementary angles
3. 1 & 9, 3 & 11, 2 & 10, 4 & 12,
   5 & 13, 7 & 15, 6 & 14, or 8 & 16
4. 4 & 9, 3 & 10, 7 & 14, or 8 & 13
5. $\angle 1, 4, 9,$ or 12: $\angle$'s 5, 8, 13 and
   16 appear to be acute but we
   don't know for certain, because
   no information is given about
   these angles.
6. no: $\overleftrightarrow{AC}$ is not parallel to $\overleftrightarrow{BD}$
7. B and C, or A and D
8. point A
9. infinite: Only two points are
   labeled, but every line contains
   an infinite number of points.
10. one: length
11. $\angle 2, \angle 3, \angle 10,$ or $\angle 11$
12. $\angle 9$

VI

$A = \dfrac{1}{2}bh = \dfrac{1}{2}(6)(2) = 6$ in^2

VII

rectangle, square,
parallelogram, rhombus

VIII

1. $\{2, 3, 4, 5, 6\}$: all elements that
   appear in either of the two sets
2. no: the element 2 is found
   in set A, but not in set B

Test 11

1. C : 4 diagonals,
   forming 5 triangles
2. A : 10 triangles
3. D : $(N - 2)180° \Rightarrow$
   $((11) - 2)180° = (9)180° = 1,620°$
4. A : pentagon
5. A : $360°$
6. B : $(N - 2)180° \Rightarrow ((8) - 2)180° =$
   $(6)180° = 1,080°$ total
   $1,080° \div 8 = 135°$ per angle
7. E : The exterior angles of a
   polygon always add up to $360°$.
8. C : $(N - 2)180° \Rightarrow ((5) - 2)180° =$
   $(3)180° = 540°$ total
   $540° \div 5 = 108°$ per angle
   $m\angle b = 180° - (36° + 108°) =$
   $180° - 144° = 36°$
9. E : $(N - 2)180° \Rightarrow ((8) - 2)180° =$
   $(6)180° = 1,080°$ total
   $1,080° \div 8 = 135°$
   $m\angle a = 135° \div 2 = 67.5°$
10. B : $(N - 2)180° \Rightarrow$
    $((6) - 2)180° = (4)180° = 720°$
11. A : $360° \div 6 = 60°$
12. C : $720°$ for all interior angles
    (from #10)
    $720° \div 6 = 120°$
    $120° \div 2 = 60°$
    for each new angle
13. D : $m\angle QVR = 180° - (30° + 120°) =$
    $180° - 150° = 30°$
14. B : $m\angle RVU = m\angle QVU - m\angle QVR =$
    $120° - 30° = 90°$
15. A : $m\angle TRU =$
    $m\angle SRQ \div 4 = 120° \div 4 = 30°$

Test 12

1. A : circumference
2. B : $360° - 50° = 310°$
3. B : diameter
4. B : tangent
5. C : they are perpendicular
6. E : inscribed in
7. D : sector
8. A : secant
9. C : arc
10. D : The measure of an inscribed angle is half that of the intercepted arc. $48° \div 2 = 24°$
11. C
12. B
13. D
14. A
15. E

Test 13

1. B : radius
2. A : circumference
3. A : πr^2
4. B : $2\pi r$
5. E : $\frac{1}{2}$ long axis $\cdot \frac{1}{2}$ short axis $\cdot \pi$
6. B : 60'
7. E : latitude
8. C : the prime meridian
9. C : $A = \pi r^2 \approx (3.14)(3^2) =$ 28.26 units^2
10. B : $C = 2\pi r \approx (2)(3.14)(3) =$ 18.84 units (radius is half the diameter)
11. A : $\frac{22}{7}$

12. C : $A = \pi r^2 \approx$ $(\frac{22}{7})(7^2) = \frac{22}{\cancel{7}} \times \frac{\cancel{49}^{7}}{1} =$ $)$ 154 in^2
13. B : $C = 2\pi r \approx (\frac{2}{1})(\frac{22}{\cancel{7}})(\frac{\cancel{7}}{1}) =$ $\frac{44}{1} = 44$ in
14. B : $A = (5)(2)(\pi) \approx (10)(3.14) =$ 31.4 in^2
15. C : $A = \pi r^2 \approx (3.14)(4^2) =$ 50.24 units^2

Test 14

1. B : the area of the base
2. E : all of the above
3. E : edges
4. C : 8 vertices
5. D : cubic units (often written as units^3)
6. A : $\pi r^2 \times h$
7. B : $V = (6)(6)(6) = 216$ in^3
8. D : faces
9. C : $V = (3)(4)(9) = 108$ units^3
10. E : $V = Bh = \pi r^2 h \approx$ $(3.14)(10^2)(6) = 1,884$ m^3
11. A : 6 faces
12. E : $V = (5)(8)(2) = 80$ m^3
13. B : $V = (10)(10)(10) = 1,000$ in^3
14. B : $V = Bh = \pi r^2 h \approx$ $(3.14)(10^2)(16) = 5,024$ ft^3
15. C : $V = Bh = \pi r^2 h \approx (3.14)(5^2)(10) =$ 785 ft^3

Test 15

1. A : triangles
2. B : altitude
3. A : slant height
4. D : prism
5. A : cylinder:

$$V = Bh = \pi r^2 h \approx$$
$$(3.14)(1.5^2)(4) = 28.26$$

cone:

$$V = \frac{1}{3}Bh = \frac{1}{3}\pi r^2 h \approx$$
$$\frac{1}{3}(3.14)(1.5^2)(4) = 9.42$$

6. D : not enough information, because we do not know the heights

7. A : sphere:

$$V = \frac{4}{3}\pi r^3 \approx \frac{4}{3}(3.14)(4^3) \approx 267.95$$

cone: $V = \frac{1}{3}Bh = \frac{1}{3}\pi r^2 h \approx$
$$\frac{1}{3}(3.14)(4^2)(4) \approx 66.99$$

8. C : same, because the spheres are the same size

9. B : cylinder: $V = Bh = (4)(10) = 40$
rectangular solid:
$V = Bh = (10)(8) = 80$
$80 > 40$

10. E : none of the above
correct formula is $V = \frac{4}{3}\pi r^3$

11. B : $V = \frac{1}{3}Bh$

12. A : $V = \frac{1}{3}Bh = \frac{1}{3}(6)(6)(10) =$
$120\ \text{in}^3$

13. E : $V = \frac{1}{3}Bh = \frac{1}{3}\pi r^2 h \approx$
$$\frac{1}{3}(3.14)(5^2)(12) = 314\ \text{in}^3$$

14. A : $V = Bh = \frac{1}{2}(3)(4)(6) = 36\ \text{ft}^3$

15. B : $V = \frac{4}{3}\pi r^3 \approx \frac{4}{3}(3.14)(6^3) =$
$904.32\ \text{m}^3$

Test 16

1. D : 6 faces
2. C : 5 faces
3. B : 4 faces
4. C : two circles and the rectangle formed by "unrolling" the side
5. D : square units
6. A : $6(7)(7) = 294\ \text{in}^2$
7. A : SA =
$2(12)(14) + 2(12)(8) + 2(8)(14) =$
$336 + 192 + 224 = 752\ \text{ft}^2$

8. A : $SA = 2\pi r^2 + 2\pi rh \approx$
$(2)(3.14)(5^2) + (2)(3.14)(5)(10) =$
$157 + 314 = 471\ \text{m}^2$

9. D : $SA = 4(\frac{1}{2}(6)(9)) + (6)(6) =$
$108 + 36 = 144\ \text{units}^2$

10. B : $SA = 2(3)(4) + 2(4)(6) + 2(3)(6) =$
$24 + 48 + 36 = 108\ \text{units}^2$

11. B : Since the square base has an area of $100\ \text{ft}^2$, it must be $\sqrt{100}$ or 10 ft on a side.
$SA = 4(\frac{1}{2}(10)(20)) + (10)(10) =$
$400 + 100 = 500\ \text{ft}^2$

12. A : $SA = 4(\frac{1}{2}(20)(30)) + (20)(20) =$
$1{,}200 + 400 = 1{,}600\ \text{m}^2$

13. D : $SA = 2(1)(3) + 2(1)(4) + 2(3)(4) =$
$6 + 8 + 24 = 38\ \text{ft}^2$

14. C: $SA = 2\pi r^2 + 2\pi rh \approx$
$2(3.14)(3^2) + 2(3.14)(3)(5) =$
$56.52 + 94.2 = 150.72 \text{ cm}^2$

15. E: "roof": $SA = 2(7)(5) = 70 \text{ m}^2$
triangles:
$$SA = 2\tfrac{1}{2}(6)(4) = 24 \text{ m}^2$$
sides:
$$SA = 2(2)(7) + 2(2)(6) = 52 \text{ m}^2$$
bottom: $SA = (6)(7) = 42 \text{ m}^2$
total:
$$SA = 70 + 24 + 52 + 42 = 188 \text{ m}^2$$

Test 17

1. C: a whole number
2. B: 6
3. B: $\sqrt{RS}$
4. E: $\sqrt{R} + \sqrt{S} = \sqrt{R} + \sqrt{S}$:
cannot be simplified
5. B: $\left(5\sqrt{X}\right)\left(6\sqrt{Y}\right) = 30\sqrt{XY}$
6. D: $10\sqrt{3}$
7. A: $\left(3\sqrt{5}\right)\left(3\sqrt{5}\right) = 9\sqrt{25} = 9(5) = 45$
8. B: $\sqrt{45} = \sqrt{9}\sqrt{5} = 3\sqrt{5}$
9. C: $\sqrt{24} = \sqrt{4}\sqrt{6} = 2\sqrt{6}$
10. E: $\sqrt{42} = \sqrt{42}$:
cannot be simplified
11. A: $\dfrac{24\sqrt{18}}{6\sqrt{9}} = \dfrac{24\sqrt{2}}{6} = \dfrac{4\sqrt{2}}{1} = 4\sqrt{2}$
12. C: $\dfrac{15\sqrt{8}}{5\sqrt{2}} = \dfrac{15\sqrt{4}}{5} = \dfrac{3\sqrt{4}}{1} =$
$3\sqrt{4} = 3(2) = 6$
13. E: cannot be simplified
14. B: $2\sqrt{3} + 3\sqrt{3} + 6\sqrt{3} =$
$(2 + 3 + 6)\sqrt{3} = 11\sqrt{3}$
15. D: $\left(5\sqrt{3}\right)\left(4\sqrt{2}\right) = 20\sqrt{6}$

Test 18

1. D: $a^2 + b^2 = c^2$
2. C: the triangle is a right triangle
3. D: $3^2 + 2^2 = H^2$
$9 + 4 = H^2$
$13 = H^2$
$\sqrt{13} = H \approx 3.61$
Of the answers given,
4 is closest.
4. C: $A^2 + B^2 = H^2$
$\sqrt{A^2 + B^2} = \sqrt{H^2}$
$\sqrt{A^2 + B^2} = H$
5. B: $3^2 + 7^2 = H^2$
$9 + 49 = H^2$
$58 = H^2$
$\sqrt{58} = H$
6. E: $4^2 + 6^2 = H^2$
$16 + 36 = H^2$
$52 = H^2$
$\sqrt{52} = H = \sqrt{4}\sqrt{13} = 2\sqrt{13}$
7. C: $6^2 + 8^2 = 10^2$
$36 + 64 = 100$
$100 = 100$: true
Since the Pythagorean
theorem applies to this
triangle, it is a right triangle.
8. B: $5^2 + 9^2 = 12^2$
$25 + 81 = 144$
$106 = 144$: not true
Since the Pythagorean
theorem does not apply to
this triangle, it is not a
right triangle.
9. C: 90°
10. A: hypotenuse

11.   D : $5^2 + L^2 = \left(\sqrt{61}\right)^2$

$$25 + L^2 = 61$$
$$L^2 = 36$$
$$L = 6$$

12.   B : $12^2 + L^2 = 13^2$

$$144 + L^2 = 169$$
$$L^2 = 25$$
$$L = 5$$

13.   A : $ST = 12$

$$12^2 + L^2 = 20^2$$
$$144 + L^2 = 400$$
$$L^2 = 256$$
$$L = 16$$

14.   B : $A = \dfrac{1}{2}bh = \dfrac{1}{2}(24)(16)$

$$= 192 \text{ units}^2$$

15.   E : $A = 8(192) = 1{,}536 \text{ units}^2$

Test 19

1.   B :   denominator
2.   E :   1
3.   A :   common denominator
4.   A : $\dfrac{5}{\sqrt{3}} = \dfrac{5\sqrt{3}}{\sqrt{3}\sqrt{3}} = \dfrac{5\sqrt{3}}{\sqrt{9}} = \dfrac{5\sqrt{3}}{3}$
5.   D : $\dfrac{8\sqrt{2}}{\sqrt{4}} = \dfrac{8\sqrt{2}}{2} = \dfrac{4\sqrt{2}}{1} = 4\sqrt{2}$
6.   B : $\dfrac{4\sqrt{3}}{\sqrt{8}} = \dfrac{4\sqrt{3}\sqrt{2}}{\sqrt{8}\sqrt{2}} = \dfrac{4\sqrt{6}}{\sqrt{16}} =$

$$\dfrac{4\sqrt{6}}{4} = \dfrac{\sqrt{6}}{1} = \sqrt{6}$$

7.   B : $\dfrac{5\sqrt{5}}{\sqrt{5}} = \dfrac{5}{1} = 5$
8.   A : $\dfrac{3\sqrt{7}}{\sqrt{10}} = \dfrac{3\sqrt{7}\sqrt{10}}{\sqrt{10}\sqrt{10}} =$

$$\dfrac{3\sqrt{70}}{\sqrt{100}} = \dfrac{3\sqrt{70}}{10}$$

9.   C : $\dfrac{4\sqrt{15}}{6\sqrt{6}} = \dfrac{4\sqrt{15}\sqrt{6}}{6\sqrt{6}\sqrt{6}} = \dfrac{4\sqrt{90}}{6\sqrt{36}} =$

$$\dfrac{4\sqrt{90}}{6(6)} = \dfrac{4\sqrt{90}}{36} = \dfrac{\sqrt{90}}{9} =$$

$$\dfrac{\sqrt{9}\sqrt{10}}{9} = \dfrac{3\sqrt{10}}{9} = \dfrac{\sqrt{10}}{3}$$

10.   D : $\dfrac{15\sqrt{11}}{\sqrt{5}} = \dfrac{15\sqrt{11}\sqrt{5}}{\sqrt{5}\sqrt{5}} = \dfrac{15\sqrt{55}}{\sqrt{25}} =$

$$\dfrac{15\sqrt{55}}{5} = \dfrac{3\sqrt{55}}{1} = 3\sqrt{55}$$

11.   C : $\dfrac{4\sqrt{3}}{\sqrt{2}} + \dfrac{2\sqrt{3}}{\sqrt{2}} = \dfrac{6\sqrt{3}}{\sqrt{2}} = \dfrac{6\sqrt{3}\sqrt{2}}{\sqrt{2}\sqrt{2}} =$

$$\dfrac{6\sqrt{6}}{\sqrt{4}} = \dfrac{6\sqrt{6}}{2} = \dfrac{3\sqrt{6}}{1} = 3\sqrt{6}$$

12.   A : $\dfrac{7}{\sqrt{5}} + \dfrac{3}{\sqrt{2}} = \dfrac{7\sqrt{5}}{\sqrt{5}\sqrt{5}} + \dfrac{3\sqrt{2}}{\sqrt{2}\sqrt{2}} =$

$$\dfrac{7\sqrt{5}}{\sqrt{25}} + \dfrac{3\sqrt{2}}{\sqrt{4}} = \dfrac{7\sqrt{5}}{5} + \dfrac{3\sqrt{2}}{2} =$$

$$\dfrac{7\sqrt{5}\,(2)}{5(2)} + \dfrac{3\sqrt{2}\,(5)}{2(5)} =$$

$$\dfrac{14\sqrt{5}}{10} + \dfrac{15\sqrt{2}}{10} = \dfrac{14\sqrt{5} + 15\sqrt{2}}{10}$$

13.   D : $\dfrac{8\sqrt{6}}{\sqrt{3}} - \dfrac{5\sqrt{3}}{\sqrt{2}} = \dfrac{8\sqrt{6}\sqrt{3}}{\sqrt{3}\sqrt{3}} - \dfrac{5\sqrt{3}\sqrt{2}}{\sqrt{2}\sqrt{2}} =$

$$\dfrac{8\sqrt{18}}{\sqrt{9}} - \dfrac{5\sqrt{6}}{\sqrt{4}} = \dfrac{8\sqrt{18}}{3} - \dfrac{5\sqrt{6}}{2} =$$

$$\dfrac{8\sqrt{9}\sqrt{2}}{3} - \dfrac{5\sqrt{6}}{2} = \dfrac{8(3)\sqrt{2}}{3} - \dfrac{5\sqrt{6}}{2} =$$

$$\dfrac{8\sqrt{2}}{1} - \dfrac{5\sqrt{6}}{2} = \dfrac{8\sqrt{2}\,(2)}{1(2)} - \dfrac{5\sqrt{6}}{2} =$$

$$\dfrac{16\sqrt{2}}{2} - \dfrac{5\sqrt{6}}{2} = \dfrac{16\sqrt{2} - 5\sqrt{6}}{2}$$

14.   E : $\dfrac{6\sqrt{11}}{\sqrt{3}} - \dfrac{2\sqrt{5}}{\sqrt{2}} = \dfrac{6\sqrt{11}\sqrt{3}}{\sqrt{3}\sqrt{3}} - \dfrac{2\sqrt{5}\sqrt{2}}{\sqrt{2}\sqrt{2}} =$

$$\dfrac{6\sqrt{33}}{\sqrt{9}} - \dfrac{2\sqrt{10}}{\sqrt{4}} = \dfrac{6\sqrt{33}}{3} - \dfrac{2\sqrt{10}}{2} =$$

$$\dfrac{2\sqrt{33}}{1} - \dfrac{\sqrt{10}}{1} = 2\sqrt{33} - \sqrt{10}$$

15.   E : $\dfrac{2\sqrt{2}}{\sqrt{8}} + \dfrac{7\sqrt{3}}{\sqrt{3}} = \dfrac{2}{\sqrt{4}} + \dfrac{7}{1} =$

$$\dfrac{2}{2} + 7 = 1 + 7 = 8$$

Unit Test II

I

1. pentagon
2. hypotenuse
3. sector
4. prism
5. rhombus
6. chord
7. sphere
8. latitude

II

$V = (10)(6)(4) = 240 \text{ in}^3$

III

Area = area of two circles plus area of "unrolled" rectangle =

$2\pi r^2 + 2\pi rh \approx$

$2(3.14)(5^2) + 2(3.14)(5)(6) =$

$157 + 188.4 = 345.4 \text{ in}^2$

IV

1. $(2\sqrt{6})(5\sqrt{10}) = (2)(5)\sqrt{6}\sqrt{10} =$
   $10\sqrt{60} = 10\sqrt{4}\sqrt{15} = 10(2)\sqrt{15} =$
   $20\sqrt{15}$

2. $3\sqrt{7} - 2\sqrt{71} + 5\sqrt{3}$:
   cannot be simplified

3. $3\sqrt{7} - 2\sqrt{7} + \dfrac{1}{2}\sqrt{7} - \dfrac{3}{2}\sqrt{7} =$
   $\left(3 - 2 + \dfrac{1}{2} - \dfrac{3}{2}\right)\sqrt{7} = \left(1 + \dfrac{-2}{2}\right)\sqrt{7} =$
   $(1 + (-1))\sqrt{7} = (0)\sqrt{7} = 0$

4. $\dfrac{\sqrt{3}}{\sqrt{6}} = \dfrac{1}{\sqrt{2}} = \dfrac{1\sqrt{2}}{\sqrt{2}\sqrt{2}} = \dfrac{\sqrt{2}}{\sqrt{4}} = \dfrac{\sqrt{2}}{2}$

V

1. $(N-2)180° \Rightarrow$
   $((6)-2)180° = (4)180° =$
   720° total
   720° ÷ 6 = 120° per angle

2. 360°: The sum of the exterior angles of a regular polygon is always 360°.

VI

1. $A = \pi r^2 \approx \left(\dfrac{22}{7}\right)\left(\dfrac{7^{\cancel{2}^7}}{1}\right) =$
   $\dfrac{154}{1} = 154 \text{ ft}^2$

2. $C = 2\pi r \approx \left(\dfrac{2}{1}\right)\left(\dfrac{22}{\cancel{7}}\right)\left(\dfrac{\cancel{7}}{1}\right) =$
   $\dfrac{44}{1} = 44 \text{ ft}$

VII

check with protractor

VIII

area of 4 triangular faces:

$A = 4\left(\dfrac{1}{2}bh\right) =$

$(4)\left(\dfrac{1}{2}\right)(4)(5) = 40 \text{ in}^2$

area of base:

$A = (4)(4) = 16 \text{ in}^2$

total area = 40 + 16 = 56 in²

IX

1. The measure of an intercepted arc is the same as the measure of the central angle that intercepts it, so m∠AXC = 82°

2. The measure of an inscribed angle is half the measure of the arc it intercepts, so
   m∠ABC = 82° ÷ 2 = 41°

X

$\text{Leg}^2 + \text{Leg}^2 = \text{Hypotenuse}^2$ or
$L^2 + L^2 = H^2$ or $A^2 + B^2 = C^2$

1. $L^2 + 6^2 = 10^2$
   $L^2 + 36 = 100$
   $L^2 = 64$
   $L = 8 \text{ ft}$

2. $L^2 + 2^2 = \left(\sqrt{13}\right)^2$

$L^2 + 4 = 13$

$L^2 = 9$

$L = 3$ units

3. $\left(2\sqrt{2}\right)^2 + \left(5\sqrt{2}\right)^2 = H^2$

$(2)(2)\sqrt{2}\sqrt{2} + (5)(5)\sqrt{2}\sqrt{2} = H^2$

$4\sqrt{4} + 5\sqrt{4} = H^2$

$4(2) + 25(2) = H^2$

$8 + 50 = H^2$

$58 = H^2$

$\sqrt{58}$ units $= H$

4. $\left(\dfrac{1}{\sqrt{2}}\right)^2 + \left(\dfrac{1}{\sqrt{3}}\right)^2 = H^2$

$\dfrac{(1)(1)}{\sqrt{2}\sqrt{2}} + \dfrac{(1)(1)}{\sqrt{3}\sqrt{3}} = H^2$

$\dfrac{1}{\sqrt{4}} + \dfrac{1}{\sqrt{9}} = H^2$

$\dfrac{1}{2} + \dfrac{1}{3} = H^2$

$\dfrac{3}{6} + \dfrac{2}{6} = H^2$

$\dfrac{5}{6} = H^2$

$\sqrt{\dfrac{5}{6}} = H$

$\dfrac{\sqrt{5}}{\sqrt{6}} = H$

$\dfrac{\sqrt{5}\sqrt{6}}{\sqrt{6}\sqrt{6}} = H$

$\dfrac{\sqrt{30}}{\sqrt{36}} = H$

$\dfrac{\sqrt{30}}{6}$ units $= H$

Test 20

1. B : hypotenuse
2. D : congruent
3. C : isosceles
4. E : Pythagorean theorem
5. B: $\sqrt{2}$
6. A : $25\sqrt{2}$
7. C : $3\sqrt{2}\sqrt{2} = 3\sqrt{4} = 3(2) = 6$
8. A : $\dfrac{9\sqrt{2}}{\sqrt{2}} = \dfrac{9}{1} = 9$
9. B : one leg =

$\dfrac{2}{\sqrt{2}} = \dfrac{2\sqrt{2}}{\sqrt{2}\sqrt{2}} = \dfrac{2\sqrt{2}}{\sqrt{4}} = \dfrac{2\sqrt{2}}{2} =$

$\dfrac{\sqrt{2}}{1} = \sqrt{2}$

both legs $= \sqrt{2} + \sqrt{2} = 2\sqrt{2}$

10. E : A, B and C
11. A : 7 because it is a
45°–45°–90° triangle
and the legs are congruent
12. C : $7\sqrt{2}$ by rule for
45°–45°–90° triangles
13. D : $m\angle \alpha = 180° - (90° + 45°) =$
$180° - 135° = 45°$
14. A : $2\sqrt{3}$ because the legs
are congruent
15. E : $2\sqrt{3}\sqrt{2} = 2\sqrt{6}$ so none
of the above

Test 21

1. D : $180° - (60° + 30°) =$
$180° - 90° = 90°$
2. A : scalene
3. D : 2 times as long
4. B : dividing by 2
5. C : $\sqrt{3}$ times as long

6. B : the side opposite the 30º angle is the short side, so the hypotenuse would be $2\left(4\sqrt{5}\right)=8\sqrt{5}$

7. E : $2(2A)=4A$

8. E : $14R\div2=7R$

9. B : $\dfrac{12}{\sqrt{3}}=\dfrac{12\sqrt{3}}{\sqrt{3}\sqrt{3}}=\dfrac{12\sqrt{3}}{\sqrt{9}}=$ $\dfrac{12\sqrt{3}}{3}=\dfrac{4\sqrt{3}}{1}=4\sqrt{3}$

10. C : $m\angle\alpha=$ $180º-\left(90º+60º\right)=$ $180º-150º=30º$

11. A : $14\div2=7$

12. D : $7\sqrt{3}$

13. B : $m\angle\beta=$ $180º-\left(90º+30º\right)=$ $180º-120º=60º$

14. A : $4\sqrt{3}$

15. C : $2(4)=8$

Test 22

1. B : unproven and obvious
2. E : postulates
3. C : congruent
4. A : 360º
5. B : rhombus
6. D : complementary
7. B : 180º
8. E : congruent
9. C : supplementary
10. B : trapezoid
11. C : $R+S>T$
12. C : The figure described may be a rhombus, rectangle or square, but is definitely a parallelogram.
13. B : right
14. E : perpendicular
15. D : parallel

Test 23

1. B : congruent
2. D : 180º
3. B : $\overline{SV}$
4. D : $\angle TVS$
5. A : $\angle VST$
6. A : $\triangle SVT$
7. E : $\overline{TV}$
8. B : $\angle SRQ$
9. D : $m\angle C=180º-118º=62º$
10. E : $m\angle B=180º-113º=67º$
11. B : $m\angle A=180º-\left(62º+67º\right)=$ $180º-129º=51º$
12. B : $m\angle D=180º-51º=129º$
13. B : $m\angle B=180º-110º=70º$
14. C : $m\angle A=180º-\left(45º+70º\right)=$ $180º-115º=65º$
15. D : $m\angle D=180º-65º=115º$

Test 24

1. B : the other two angles
2. A : they are congruent by SSS
3. E : I, II, and IV, because of SAS
4. A : $A=A$
5. C : $\overline{GH}\cong\overline{HE}$
6. D : congruent
7. C : SSS
8. B : SAS
9. A : definition of a rhombus
10. D : reflexive property
11. C : SSS
12. D : definition of midpoint
13. C : vertical angles are congruent
14. B : SAS
15. B : postulates are unproven statements used to prove theorems

Test 25

1. E : AAA, because sides may be different lengths
2. D : They are congruent.
3. A : prove the triangles congruent
4. C : the midpoint
5. B : If one set of corresponding sides are congruent, the triangles may be proved congruent by ASA or AAS.
6. A : $\angle JKZ \cong \angle XKZ$
7. C : parallelogram
8. E : ASA
9. B : congruent
10. A : reflexive property
11. D : definition of a bisector
12. B : $\overline{RV} \cong \overline{TV}$
13. E : definition of a parallelogram
14. C : alternate interior angles
15. A : ASA

Test 26

1. E : one congruent angle is already given
2. B : Pythagorean theorem
3. D : SSS
4. A : SAS
5. B : AAS
6. B : HA
7. C : definition of a rectangle
8. B : opposite sides of a rectangle are congruent (APT)
9. C : reflexive property
10. A : HL
11. E : definition of a midpoint
12. B : $\overline{MQ} \cong \overline{MQ}$
13. D : LL
14. A : $\overline{PM} \cong \overline{RM}$: CPCTRC

    For #14 and #15, you may not assume the figure is a rectangle.

15. B : $\angle NRQ$ is a right angle: all others may be proved congruent with CPCTRC.

Test 27

1. C : have the same shape but not the same size
2. A : corresponding sides are congruent
3. B : two sets of congruent angles
4. C : ratio of short legs is also $\frac{1}{3}$
5. D : similar
6. A : $\frac{8}{10} = \frac{4}{5}$
7. D : they are similar:

    $\frac{3}{6}$, $\frac{5}{10}$, and $\frac{6}{12}$ all $= \frac{1}{2}$

8. C : two congruent angles proves similarity, not congruence
9. E : $\frac{2}{10} = \frac{1}{5}$
10. E : reflexive property
11. B : perpendicular lines form right angles
12. A : $\triangle XSY$ and $\triangle RSQ$ are similar by AA
13. E : vertical angles
14. A : alternate interior angles
15. C : AA postulate

Test 28

1. C : moving and changing shapes on a grid
2. A : translation
3. E : reflection
4. B : dilation
5. C : flipped
6. A : rotation
7. E : counterclockwise
8. A : degrees
9. C : reflection

10. B : translation of 5 spaces
11. C : rotation of 90° around the origin
12. A : R
13. E : none
14. D : V; Each point on figure Q has been moved to the left 5 and up 2.
15. B : S; Q has been translated and rotated, so its transformation includes rotation.

Test 29

1. A : triangles
2. C : right
3. D : cosine
4. A : tangent
5. E : none of the above
6. B : $\frac{B}{C}$
7. B : $\frac{A}{B}$
8. E : $\frac{A}{C}$
9. A : $\frac{5\sqrt{3}}{10} = \frac{\sqrt{3}}{2}$
10. A : $\frac{5\sqrt{3}}{10} = \frac{\sqrt{3}}{2}$
11. D : $\frac{5}{5\sqrt{3}} = \frac{1}{\sqrt{3}} = \frac{\sqrt{3}}{\sqrt{3}\sqrt{3}} = \frac{\sqrt{3}}{\sqrt{9}} = \frac{\sqrt{3}}{3}$
12. C : 30°
13. C : $\frac{4}{5}$
14. A : $\frac{3}{5}$
15. B : $\frac{3}{4}$

Test 30

1. B : cosecant
2. D : secant
3. B : cotangent
4. E : none of the above: it is the cotangent
5. A : secant
6. C : cosecant
7. D : $\frac{C}{A}$
8. B : $\frac{A}{C}$
9. A : $\frac{C}{A}$
10. C : $\frac{2\sqrt{3}}{2} = \frac{\sqrt{3}}{1} = \sqrt{3}$
11. E : $\frac{4}{2} = 2$
12. B : $\frac{4}{2} = 2$
13. D : $\sin^2\theta + \cos^2\theta = 1$
14. E : $\frac{4\sqrt{2}}{4} = \frac{\sqrt{2}}{1} = \sqrt{2}$
15. C : $\frac{4}{4} = 1$

Unit Test III
I

1. axiom or postulate
2. dilation
3. reflection
4. tangent
5. secant
6. similar
7. sphere
8. cotangent

II

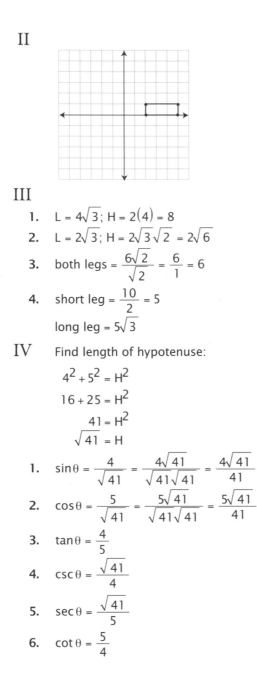

III

1. $L = 4\sqrt{3}; H = 2(4) = 8$

2. $L = 2\sqrt{3}; H = 2\sqrt{3}\sqrt{2} = 2\sqrt{6}$

3. both legs $= \dfrac{6\sqrt{2}}{\sqrt{2}} = \dfrac{6}{1} = 6$

4. short leg $= \dfrac{10}{2} = 5$

    long leg $= 5\sqrt{3}$

IV   Find length of hypotenuse:

$$4^2 + 5^2 = H^2$$
$$16 + 25 = H^2$$
$$41 = H^2$$
$$\sqrt{41} = H$$

1. $\sin\theta = \dfrac{4}{\sqrt{41}} = \dfrac{4\sqrt{41}}{\sqrt{41}\sqrt{41}} = \dfrac{4\sqrt{41}}{41}$

2. $\cos\theta = \dfrac{5}{\sqrt{41}} = \dfrac{5\sqrt{41}}{\sqrt{41}\sqrt{41}} = \dfrac{5\sqrt{41}}{41}$

3. $\tan\theta = \dfrac{4}{5}$

4. $\csc\theta = \dfrac{\sqrt{41}}{4}$

5. $\sec\theta = \dfrac{\sqrt{41}}{5}$

6. $\cot\theta = \dfrac{5}{4}$

V

$$(X + 4) + (2X + 6) + (-5X) = 12$$
$$X + 2X + (-5X) + 4 + 6 = 12$$
$$-2X + 10 = 12$$
$$-2X = 2$$
$$X = -1$$

$X + 4 \Rightarrow (-1) + 4 = 3$

$2X + 6 \Rightarrow 2(-1) + 6 = -2 + 6 = 4$

$-5X \Rightarrow -5(-1) = 5$

Sides are 3, 4 and 5.

$$3^2 + 4^2 = 5^2$$
$$9 + 16 = 25$$
$$25 = 25: \text{true}$$

Since the Pythagorean theorem applies, this is a right triangle.

VI   Please note: The proofs given here may not be the only valid options. As long as each statement is based on given information, valid postulates, definitions and theorems, or on statements made previously within the proof, the student's proof can be considered correct.

1.

$\overline{AC}$ bisects $\angle BAD$	given
$\angle BCA \cong \angle DCA$	given
$\angle BAC \cong \angle DAC$	a bisector divides the angle into equal parts
$\overline{AC} \cong \overline{AC}$	reflexive property
$\angle BCA$ is a right angle	given
$\angle DCA$ is a right angle	supplementary angles
$\triangle BAC \cong \triangle DAC$	LA

2.

$\overline{BD} \parallel \overline{CE}$	given
$\angle ABD \cong \angle ACE$	corresponding angles
$\angle BAD \cong \angle CAE$	reflexive property
$\triangle ACE \sim \triangle ABD$	AA

3.

$m\angle ADC = m\angle BCD = 90°$	definition of a rectangle
$\overline{DC} \cong \overline{DC}$	reflexive property
$\overline{AD} \cong \overline{BC}$	opposite sides of a rectangle are congruent (APT)
$\triangle ADC \cong \triangle BCD$	SAS or LL

Final Exam

I

1. cosine
2. obtuse
3. arc
4. complementary
5. plane
6. trapezoid
7. cube
8. collinear
9. congruent
10. perimeter

II

1. trapezoid
2. $\angle 12$
3. $m\angle 6 = m\angle 8 = 60°$
   corresponding angles
4. $m\angle 5 = 180° - (m\angle 4 + m\angle 6) =$
   $180° - (60° + 90°) =$
   $180° - 150° = 30°$
5. $\triangle BDC$ is a $30°-60°-90°$ triangle
   hypotenuse = 8 in
   $\overline{BD}$ (short leg) $= 8 \div 2 = 4$ in
   $\overline{BC}$ (long leg) $= 4\sqrt{3}$

6. $m\angle 14 = 180° - m\angle 5 =$
   $180° - 30° = 150°$
7. no, line EC is not parallel to line AC
8. point E
9. Let X = length of $\overline{AE}$
   $$\frac{20}{8} = \frac{X}{4}$$
   $8X = (4)(20)$
   $8X = 80$
   $X = 10$
10. First find length of $\overline{AC}$:
    $\triangle EAC$ is a $30°-60°-90°$ triangle, so the long leg is $\sqrt{3}$ times the short leg or $10\sqrt{3}$
    $AB = AC - BC = 10\sqrt{3} - 4\sqrt{3} = 6\sqrt{3}$

III

1.

$\overline{CE} \cong \overline{CA}$	given
$\angle ABC \cong \angle CDE$	given
$\angle ACB \cong \angle DCE$	vertical angles
$\triangle ABC \cong \triangle CDE$	AAS

2.

$\overline{AB} \cong \overline{BC}$	given
$\angle BEC$ is a right angle	given
$\angle BEA$ is a right angle	supplementary
$\overline{BE} \cong \overline{BE}$	reflexive property
$\triangle ABE \cong \triangle CBE$	HL
$\overline{AE} \cong \overline{CE}$	CPCTRC

IV.

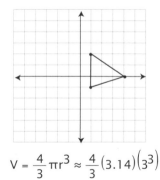

V

$V = \frac{4}{3}\pi r^3 \approx \frac{4}{3}(3.14)(3^3)$

$= 113.04$ cm³

If the fractional value of π is used, the answer would be 113.14 cm³.

VI $SA = 2(2)(5) + 2(2)(7) + 2(5)(7) =$

$20 + 28 + 70 = 118 \text{ cm}^2$

VII 360° total of all angles

360° ÷ 45° = 8 sides; octagon

VIII

1. $\left(3\sqrt{2}\right)\left(4\sqrt{22}\right) = (3)(4)\sqrt{2}\sqrt{22} =$

$12\sqrt{44} = 12\sqrt{4}\sqrt{11} = 12(2)\sqrt{11} =$

$24\sqrt{11}$

2. $\dfrac{4}{\sqrt{3}} - \dfrac{2\sqrt{6}}{\sqrt{2}} = \dfrac{4\sqrt{3}}{\sqrt{3}\sqrt{3}} - \dfrac{2\sqrt{3}}{1} =$

$\dfrac{4\sqrt{3}}{\sqrt{9}} - \dfrac{2\sqrt{3}}{1} = \dfrac{4\sqrt{3}}{3} - \dfrac{2\sqrt{3}}{1} =$

$\dfrac{4\sqrt{3}}{3} - \dfrac{2\sqrt{3}\,(3)}{1(3)} = \dfrac{4\sqrt{3}}{3} - \dfrac{6\sqrt{3}}{3} =$

$\dfrac{4\sqrt{3} - 6\sqrt{3}}{3} = \dfrac{-2\sqrt{3}}{3}$

3. $-3\sqrt{5} + \sqrt{5} = (-3+1)\sqrt{5} = -2\sqrt{5}$

4. $\sqrt{2} + \sqrt{3} + \sqrt{4} + \sqrt{1} =$

$\sqrt{2} + \sqrt{3} + 2 + 1 = \sqrt{2} + \sqrt{3} + 3$

IX $C = \pi d \Rightarrow \quad 8\pi = \pi d$

$\dfrac{8\pi}{\pi} = \dfrac{\pi d}{\pi}$

$8 = d$

radius $= (\dfrac{1}{2})8 = 4$

X Check with ruler:
smaller segments should
each measure 2 inches.

XI The measure of a central angle
is equal to the measure of the
arc it intercepts.
m∠AXC = 98°
The measure of an inscribed
angle is half the measure of the
arc it intercepts.
m∠ABC = 98° ÷ 2 = 49°

XII $L^2 + 2^2 = 5^2$

$L^2 + 4 = 25$

$L^2 = 21$

$L = \sqrt{21}$

XIII Start by drawing a diagram.

Sine is $\dfrac{3}{5} = \dfrac{\text{opposite}}{\text{hypotenuse}}$

so we know that the hypotenuse
is 5, and one leg is 3.

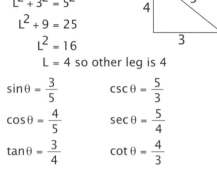

$L^2 + 3^2 = 5^2$

$L^2 + 9 = 25$

$L^2 = 16$

L = 4 so other leg is 4

$\sin\theta = \dfrac{3}{5}$ $\qquad \csc\theta = \dfrac{5}{3}$

$\cos\theta = \dfrac{4}{5}$ $\qquad \sec\theta = \dfrac{5}{4}$

$\tan\theta = \dfrac{3}{4}$ $\qquad \cot\theta = \dfrac{4}{3}$

Symbols & Tables

SYMBOLS

<	less than
>	greater than
≤	less than or equal to
≥	greater than or equal to
=	equal in numerical value
≠	not equal
≈	approximately equal
≅	congruent
~	similar
$\sqrt{\ }$	square root (radical sign)
π	pi $\left(\approx 3.14 \text{ or} \approx \dfrac{22}{7}\right)$
α	alpha
β	beta
γ	gamma
δ	delta

θ	theta
{ }	set
⊂	subset
∩	intersection
∪	union
∅	empty set
∞	infinity
↔	line
→	ray
—	line segment
⌒	arc
∠	angle
m∠	measure of angle
⊥	perpendicular
∥	parallel
∟	right angle

VOLUME

Measure of inscribed angle =
½ measure of intercepted arc

Perimeter: add the length of each side

Circumference of a circle = $2\pi r$ or πd

Area

rectangle = bh (or base x height)
triangle = ½ bh
square = bh or s^2
parallelogram or rhombus = bh
trapezoid = $\dfrac{base_1 + base_2}{2}(h)$
circle = πr^2
ellipse =
(½ short axis)(½ long axis)(π)

Surface Area

rectangular solid, cube, prism, pyramid: add the area of each face
cylinder: 2(area of base) + $2\pi rh$
    or $2\pi r^2 + 2\pi rh$
sphere: $4\pi r^2$

Volume

(B = area of base)
rectangular solid, prism = Bh
cylinder = Bh
pyramid and cone = $(^1/_3)(Bh)$
sphere: $\frac{1}{3}\pi r^3$

MISCELLANEOUS

Number of degrees

sum of interior angles of a regular polygon

$(N - 2) \times 180^0$ (N = number of sides)

sum of interior angles of a quadrilateral: 360^0

sum of interior angles of a triangle: 180^0

sum of eterior angles of a regular polygon: 360^0

Measure of central angle = measure of intercepted arc

Pythagorean Theorem

$$L^2 + L^2 = H^2$$

$$\sin^2\theta + \cos^2\theta = 1$$

Special Triangles

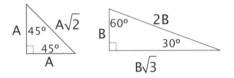

Trigonometry functions

$$\text{sine}\left(\sin\right) = \frac{\text{opposite}}{\text{hypotenuse}}$$

$$\text{cosine}\left(\cos\right) = \frac{\text{adjacent}}{\text{hypotenuse}}$$

$$\text{tangent}\left(\tan\right) = \frac{\text{opposite}}{\text{adjacent}}$$

$$\text{cosecant}\left(\csc\right) = \frac{\text{hypotenuse}}{\text{opposite}}$$

$$\text{secant}\left(\sec\right) = \frac{\text{hypotenuse}}{\text{adjacent}}$$

$$\text{cotangent}\left(\cot\right) = \frac{\text{adjacent}}{\text{opposite}}$$

Glossary

A–B

Acute angle - has a measure greater than 0° and less than 90°

Acute triangle - each angle is less than 90°

Adjacent angles - share common side and have the same origin

Apothem - perpendicular line from side to center of a regular polygon

Arc - a piece of the circumference of a circle

Area - the number of square units in a two-dimensional figure

Associative property - the way terms are grouped does not affect the answer; true for addition and multiplication

Axiom - same as a postulate, known to be true but cannot be proven

Bearing - the compass direction or angle used to navigate a ship or plane in the desired direction

Bisect - to cut into two equal parts

Bisector - a line or line segment that divides another line segment or an angle into two equal parts

C

Carroll diagram - a chart used to organize what is known and discover what is unknown in simple logic problems

Cartesian coordinate system - the X-Y graphing system named for René Descartes

Chord - line drawn between two points on the circumference of a circle

Circumference - the distance around the outside of a circle

Collinear - refers to two or more points on the same line

Commutative property - the order of terms does not affect the answer; true for addition and multiplication

Compass rose - a design used on maps to indicate directions

Complementary angles - any two angles whose measures add up to 90°

Conclusion - in formal logic, the final statement based on the major and minor premises

Conditional - the major premise of a logical argument or syllogism

Converse - in formal logic, a statement formed by reversing the order of the major premise

Cone - a three dimensional figure rising to a point from a circular base, as in ice cream cone

Congruent - having the same size and shape

Converse - the reverse of a postulate or theorem

Coplanar - refers to two or more lines lying in the same plane

Cosecant - in trigonometry, the inverse of the sine

Cosine - trigonometric function defined as adjacent angle over the hypotenuse

Cotangent - in trigonometry, the inverse of the tangent

CPCTRC - corresponding parts of congruent triangles are congruent

Cube - a rectangular solid with all the edges having the same length

Cylinder - three dimensional figure with a circular base and top, and perpendicular sides, as in a can

D–E

Decagon - regular polygon with ten sides and angles

Deductive reasoning - starts with a known fact and reaches a conclusion using the rules of logic

Diameter - a line drawn from one side of a circle to another that goes through the center of the circle

Dilation - a figure is enlarged or reduced without changing its shape

Dodecagon - regular polygon with twelve sides and angles

Ellipse - a "stretched" circle with two centers or foci

Empty set - no possible answer; same as the null set

Equiangular triangle - all angles are congruent, angular means angle

Equilateral triangle - all sides are congruent, lateral means side

Exterior angles - the angles on the outside formed by a transversal intersecting parallel lines, or the outside angles formed by extending the sides of a polygon

F–H

Face - the flat surface of a three-dimensional figure or solid

Factors - the numbers being multiplied to find the product; in the geometric illustration, the dimensions of a rectangle

Formal logic - uses deductive reasoning and predetermined rules to reach a conclusion

Golden rectangle - a rectangle whose proportions are thought to be particularly pleasing; ratio between the sides is the golden ratio

Hexagon - regular polygon with six sides and angles

Hypotenuse - the side opposite the right angle in a right triangle

I–L

Inductive reasoning - comes to a specific conclusion based on general observations

Inscribed polygon - a polygon drawn inside another figure, so that each vertex of the inscribed polygon touches the outer figure

Intersection - where two or more figures or sets overlap

Interior angles - the angles on the inside formed by a transversal intersecting parallel lines, or inside angles of a polygon

Intersection - where two or more lines or planes meet, where two lines share a common point or two planes share a common line, or where two sets overlap

Irrational numbers - cannot be written as a rational number; if changed to a decimal the numbers continue ...

Isosceles triangle - a triangle with two congruent sides

Latitude - horizontal lines that measure the north-south distance from the equator

Line - an infinite number of connected points

Line segment - a finite or measurable piece of a line

Longitude - vertical lines extending from pole to pole that measure the east-west distance from the prime meridian

M–O

Major premise - the first statement in a syllogism; also called the conditional

Minor premise - the second statement in a syllogism

Mobius strip - a strip of paper joined in such a way that one continuous line can be drawn on both sides

Midpoint - marks the center of a line segment

Null set - set with no members; empty set

Oblique prism - a prism whose sides are not at right angles to the bases

Obtuse angle - has a measure greater than 90° and less than 180°

Obtuse triangle - one angle greater than 90°

Octagon - regular polygon with eight sides and angles

Origin - the point where the X and Y axes intersect on a graph, or the starting point of a ray

P

Parallel lines - two straight lines in the same plane that don't intersect

Parallelogram - a four-sided figure with opposite sides parallel and of equal length; angles may or may not be right angles

Pentagon - regular polygon with five sides and angles

Perimeter - the distance around the outside of a two-dimensional figure

Perpendicular bisector - a line or line segment that is perpendicular to another line segment through its midpoint

Perpendicular lines - two lines that form right angles where they intersect

Pi - Greek letter π, the relationship between the diameter and circumference of a circle, equals 22/7, approximately 3.14

Plane - infinite number of connected lines lying in the same flat surface; has length and width; two-dimensional

Plane geometry - study of two-dimensional figures

Point - in geometry, the smallest possible unit of measure; has position but no dimension

Polygon - a two-dimensional shape with three or more sides; a regular polygon has all sides equal

Polynomial - an algebraic expression with more than one term

Postulate - assumed to be true, but cannot be proven; an observation

Prism - a solid with two parallel bases and lateral surfaces that are parallelograms

Protractor - device used to measure number of degrees in an angle

Pyramid - has a square base and four triangular faces, or a triangular base and three triangular faces

Pythagorean theorem - describes the relationship between the legs and the hypotenuse of a right triangle
$$(L^2 + L^2 = H^2)$$

Q–R

Quadrilateral - a four-sided polygon

Radical - a numeral written with a square root sign

Radius - a line from the center of a circle to the edge; plural is radii

Rational number - the result of dividing two whole numbers; can be a whole number, decimal or fraction

Ratio - relationship between two numbers expressed as a fraction

Rectangle - means "right angle" - a four-sided polygon with four right angles and opposite sides the same length

Ray - part of a line having a definite starting point and proceeding in one direction only

Rectangular solid - has six faces and all the angles are 90°

Reflection - a figure is flipped in a line to form a mirror image

Reflex angle - has a measure between 180° and 360°

Reflexive property - A = A

Regular polygon - a polygon with all sides and all angles congruent

Remote interior angle - interior angle farthest from the one under discussion

Rhombus - a quadrilateral with four sides congruent

Right angle - an angle with a measure of 90°

Right triangle - a triangle with one 90° angle

Rotation - a geometric figure is rotated around a given point

S

Scalene triangle - has no congruent sides

Secant - a line intersecting the circumference of a circle in two places, or in trigonometry, the inverse of the cosine

Sector - pie-shaped section of a circle

Set - a collection of numbers or things

Similar - two geometric figures whose angles are congruent, and whose sides have the same ratio

Sine - trigonometric function defined as opposite angle over the hypotenuse

Solid geometry - study of three-dimensional figures

Sphere - a three-dimensional circle; "ball"

Square - a quadrilateral with four congruent sides and four right angles (also congruent); a four-sided regular polygon

Square roots - the factors of a square, 3 is the square root of 9

Straight angle - 180° angle

Subset - a part of a given set

Supplementary angles - any two angles whose measures add up to 180°

Surface area - the sum of the areas of all the faces of a solid

Syllogism - the reasoning process in formal logic

Symmetry - if A = B, then B = A

T–U

Tangent - (geometry) a line that intersects the circumference of a circle at exactly one point

Tangent - (trigonometry) the function defined as opposite side over the adjacent side

Theorem - a statement that can be proven true by the use of postulates

Transformational geometry - concerned with moving and changing figures on a graph

Transitive property - If A = B and B = C, then A = C

Translation - a figure is moved, but not changed

Transversal - a line that intersects two or more lines

Trapezoid - a four-sided polygon with two parallel sides and two sides that are not parallel

Triangle - polygon with three sides

Trigonometry - study of the measurements of triangles and their relationships

Union - two or more sets combined

V–Z

Vector - a measurement involving both direction and speed or force

Venn diagram - two or three overlapping circles or ovals; used to illustrate intersection and union of sets

Vertex - the point of an angle where the lines, segments, or rays intersect

Vertical angles - opposite angles formed by the intersection of two lines

Volume - the number of cubic units in a three-dimensional figure

Secondary Levels Master Index

This index lists the levels at which main topics are presented in the instruction manuals for Pre-Algebra through PreCalculus. For more detail, see the description of each level at www.mathusee.com.

Absolute value .... Pre-Algebra, Algebra 1
Additive inverse .... Pre-Algebra
Age problems .... Algebra 2
Angles .... Geometry, PreCalculus
Angles of elevation & depression ....PreCalc.
Arc functions .... PreCalculus
Area.. Geometry
Associative property .... Pre-Algebra,
　Algebra 1
Axioms .... Geometry
Bases other than 10 .... Algebra 1
Binomial theorem........................ Algebra 2
Boat in current problems.............. Algebra 2
Chemical mixtures .... Algebra 2
Circumference .... Geometry
Circumscribed figures................. Geometry
Cofunctions .... PreCalculus
Coin problems ....Algebra 1 & 2
Commutative property .... Pre-Algebra,
　Algebra 1
Completing the square................ Algebra 2
Complex numbers .... Algebra 2
Congruency .... Pre-Algebra, Geometry
Conic sections ....Algebra 1 & 2
Conjugate numbers .... Algebra 2
Consecutive integers..............Algebra 1 & 2
Determinants .... Algebra 2
Difference of two squares ......Algebra 1 & 2
Discriminants............................... Algebra 2
Distance formula .... Algebra 2
Distance problems .... Algebra 2
Distributive property............... Pre-Algebra,
　Algebra 1
Expanded notation.................... Pre-Algebra
Exponents
　fractional........................Algebra 1 & 2
　multiply & divide ....Algebra 1 & 2
　negative ....Algebra 1 & 2
　notation .... Pre-Algebra
　raised to a power.............Algebra 1 & 2
Factoring polynomials............Algebra 1 & 2
Functions.................................. PreCalculus
Graphing
　Cartesian coordinates .... Algebra 1
　circle ....Algebra 1 & 2

ellipse ....Algebra 1 & 2
hyperbola........................Algebra 1 & 2
inequality........................Algebra 1 & 2
line.................................Algebra 1 & 2
parabola.........................Algebra 1 & 2
polar .... PreCalculus
trig functions.................... PreCalculus
Greatest common factor........... Pre-Algebra
Identities .... PreCalculus
Imaginary numbers .... Algebra 2
Inequalities .... Algebra 1 & 2, PreCalculus
Inscribed figures .... Geometry
Integers .... Pre-Algebra
Interpolation .... PreCalculus
Irrational numbers .... Pre-Algebra
Latitude and longitude .... Geometry
Least common multiple............. Pre-Algebra
Limits .... PreCalculus
Line
　graphing.........................Algebra 1 & 2
　properties of .... Geometry
Logarithms .... PreCalculus
Matrices....................................... Algebra 2
Maxima & minima .... Algebra 2
Measurement, add & subtract ... Pre-Algebra
Metric-Imperial conversions ...Algebra 1 & 2
Midpoint formula .... Algebra 2
Motion problems.......................... Algebra 2
Multiplicative inverse .... Pre-Algebra
Natural logarithms .... PreCalculus
Navigation .... PreCalculus
Negative numbers...................... Pre-Algebra
Number line............. Pre-Algebra, Algebra 1
Order of operations .... Pre-Algebra,
　Algebra 1
Parallel & perpendicular lines
　graphing.........................Algebra 1 & 2
　properties of .... Geometry
Pascal's triangle .... Algebra 2
Percent problems........................ Algebra 2
Perimeter.................................... Geometry
Pi .... Pre-Algebra, Geometry
Place value................................ Pre-Algebra
Plane .... Geometry
Points, lines, rays....................... Geometry

Polar coordinates & graphs .... PreCalculus
Polygons
      area.. Geometry
      similar............... Pre-Algebra, Geometry
Polynomials
      add.................... Pre-Algebra, Algebra 1
      divide .... Algebra 1
      factor ....Algebra 1 & 2
      multiply .... Pre-Algebra, Algebra 1 & 2
Postulates .... Geometry
Prime factorization................... Pre-Algebra
Proofs.. Geometry
Pythagorean theorem.............. Pre-Algebra,
      Geometry, PreCalculus
Quadratic formula........................ Algebra 2
Radians.................................... PreCalculus
Radicals .... Pre-Algebra,
      Algebra 1 & 2, Geometry
Ratio & proportion .... Pre-Algebra,
      Geometry, Algebra 2
Rational expressions.................... Algebra 2
Real numbers........................... Pre-Algebra
Reference angles....................... PreCalculus
Same difference theorem .... Pre-Algebra
Scientific notation ....Algebra 1 & 2
Sequences & series .... PreCalculus
Sets .... Geometry
Significant digits .... Algebra 1
Similar polygons ...... Pre-Algebra, Geometry
Simultaneous equations ....Algebra 1 & 2
Sine, cosine, & tangent, laws of. PreCalculus
Slope-intercept formula..........Algebra 1 & 2
Solve for unknown .... Pre-Algebra,
      Algebra 1 & 2
Special triangles....... Geometry, PreCalculus
Square roots ...... Pre-Algebra, Algebra 1 & 2
Surface area............. Pre-Algebra, Geometry
Temperature conversions.......... Pre-Algebra
Time
      add & subtract.................... Pre-Algebra
      military............................... Pre-Algebra
Time & distance problems............ Algebra 2
Transformations .... Geometry
Transversals .... Geometry
Triangles .... Geometry, PreCalculus
Trig ratios............... Geometry, PreCalculus
Unit multipliers ....Algebra 1 & 2
Vectors .... Algebra 2, PreCalculus
Volume.................... Pre-Algebra, Geometry
Whole numbers.......................... Pre-Algebra

Algebra Review Topics
in Geometry Student Text

Applications with geometry..26E–30E
Area and perimeter with unknowns 9E
Binomials (multiplying)... 16E
Coefficients (factoring with) 20E
Commutative property 3E
Difference of two squares 22E
Distributive property.. 5E
Elimination 13E
Equations
     parallel lines... 10E
     perpendicular lines.. 10E
     slope intercept form of a line.......................... 8E
     solving 4E, 12E, 13E, 23E
     standard form of a line 11E
Exponents..15E, 19E
Factoring
     polynomials... 17E, 18E, 20E
     repeated 22E
     solving with 23E
     trinomials 17E; 18E; 20E
Fractional exponents... 19E
Line
     equations 8E, 10E, 11E
     parallel and perpendicular 10E
     slope 7E
Negative numbers...6E, 18E
Order of operations 2E
Polynomials 17E, 18E, 20E, 23E
Pythagorean theorem with unknowns............................ 21E
Radical sign 14E
Repeated factoring.. 22E
Scientific notation 24E, 25E
Simultaneous equations 12E, 13E
Slope intercept equation 8E
     Slope of a line.. 7E
Solving for an unknown
     elimination 13E
     simple equations 4E
     substitution.. 12E
     two unknowns...12E, 13E
     with factoring 23E
Square roots 14E
Squaring negative numbers 6E
Standard equation of a line 11E
Substitution 12E
Y-intercept... 8E

Geometry Index

Topic..**Lesson**

AA postulate...27
AAS postulate 25
Altitude 15
Amplified Parallelogram Theorem (APT) ..25
Angles
     acute 4
     adjacent 6
     alternate exterior.............................7
     alternate interior.............................7
     bisecting an angle 5
     central 12
     complementary.....................6, 7H, 8H
     corresponding 7
     inscribed 12
     measuring 3, 4
     obtuse...4
     reflex 4
     remote interior 23, 24H
     right 4
     straight 4
     supplementary.................................6
     vertical 6, 6H, 28H
Apothem.................................student 18C
Archimedes...17H
Arc of a circle..12
Area
     circle 13, 16H
     irregular polygon...........................13H
     rectangle 9, 9H
     ring 21H
     trapezoid....................................9, 9H
     triangle................................9, 9H, 14H
Area/perimeter relationships 9H
ASA postulate 25
Associative property 3H
Axioms 22
Bearings..4H
Bisector 5
Carroll diagram.......................................1H
Chord 12
Circle
     area...13, 16H
     circumference...........................12,13
     inscribed and circumscribed 12, 12H
     parts of 12
     proofs 27H, 28H
Collinear..1

Topic..**Lesson**

Commutative property 3H
Compass rose...4H
Complementary angles 6, 7H, 8H
Conditional...23H
Cone...15
Congruent 1, 24, 25, 26
Construction....................5H, 6H, 12H
Converse...23H
Corresponding parts 23, 25
Cube
     surface area...................................16
     volume 14
Cylinder
     surface area.................16, 17H, 19H
     volume............14, 14H, 15H, 17H, 19H
Deductive reasoning 22H
Diameter..12
Dilation...28
Ellipse 12, 13
Empty set...2
Exterior angles
     alternate...7
     of a polygon 11
Faces 14, 15, 16
Geometric proofs . 24, 25, 26, 27, 27H, 28H
Golden rectangle and ratio..............10H HA
     postulate 26
Hexagon...12H
HL postulate 26
Hypotenuse 18, 20, 21, 29, 30
Inductive reasoning...............................22H
Inscribed figures 12, 7H, 12H, 28H
Intersection 2
Intercepted arc.......................................28H
Interior angles
     polygons 11
     triangles 8
     with transversals 7
     LA postulate 26
     Latitude 13
     Lines 1, 5, 7
     Line segment 1
     LL postulate 26
Logic
     formal 22H, 23H
     with charts 1H, 11H
Longitude 13

Topic	Lesson
Major and minor premises	22H
Midpoint	5
Mobius strip	2H
Null set	2
Oblique prism	19H
Octagon	7H, 12H
Order of operations	2H
Parallel lines	5, 7
Parallelogram	
area	9
perimeter	8
Perimeter of quadrilateral	8
Perimeter/area relationships	9H
Perpendicular lines	5
Plane	2
Plane geometry	2
Points	1
Polygons	11, 13H
Postulates	7, 22
Premise	22H
Prism	15
Proofs	24, 25, 25H, 26, 26H, 27, 27H, 2H8
Pyramid	
surface area	16
volume	15
Pythagorean theorem	18, 18H, 21H, 30
Quadrilaterals	
area	9
perimeter	8
Radicals	
add and subtract	17
multiply and divide	17
simplify	17, 19
Radius	12
Ratio	27
Ray	1
Rectangle	
area	9
perimeter	8
Rectangular solid	
surface area	16, 16H
volume	14, 15H, 16H
Reflection	28
Regular polygons	
area	student 18
exterior angles	11
interior angles	11
types	11
Rhombus	8, 9
Right triangles	
defined	10
proving them congruent	26
Pythagorean theorem	18
special triangles	20, 21
Ring	21H
Rotation	28
SAS postulate	24
Secant	
of a circle	12
trig ratio	30
Sector of a circle	12
Sets and set notation	2, 2H, 3H
Similar triangles	27
Special triangles	
30°-60°-90°	21
45°-45°-90°	20
Sphere	
defined	12
surface area	17H, student 21
volume	15, 15H, 17H
Square	
area	9
perimeter	8
Square root	17
SSS postulate	24
Surface area	16, 16H, 19H
Syllogism	22H
Tangent	
of a circle	12, 8H, 27H, 28H
trig ratio	29, 29H, 30
Theorem	22
Transversal	7, 7H
Trapezoid	
area	9, 9H
perimeter	8
Transformational geometry	28
Triangle	
acute	10
area	9
congruent	24, 25, 26
constructing	10
corresponding parts	23
equiangular	10

Topic .... **Lesson**

    equilateral 10
    interior angles add to 180° 8
    isosceles 10
    limitations of 10
    obtuse 10
    right 10, 18
    scalene 10
    special 20, 21
Trigonometry
    secant, cosecant, cotangent 30
    sine, cosine, tangent 29, 29H, 30H
Union 2
Vector 20H
Venn diagram 2H, 3H
Vertex 3, 14, 15
Volume
    cone 15
    cube 14
    cylinder 14, 14H, 15H, 17H, 19H
    prism 15, 19H
    pyramid 15
    rectangular solid 14, 15H, 16H
    sphere 15, 15H, 17H, 19H